Praise for Coconut Delights

A Collection of Coconut Recipes
Cookbook Delights Series – Book 4

"***Coconut Delights Cookbook*** is not only a cookbook, but a wealth of information about coconut. It includes fascinating facts, folklore, history of coconut, health and nutrition, types of coconut, and even coconut-themed poetry.

In addition to all this information it has a collection of over 270 recipes that are delicious, and will be enjoyed by your family and friends.

This is a great value for the price and makes a wonderful gift."…

Dr. James G. Hood
Editor

"One of the most versatile foods in the world now has a book devoted to revealing its entire splendor. The ***Coconut Delights Cookbook*** is brimming with hundreds of recipes, facts, and folklore regarding one of the world's most useful natural provisions.

Author Karen Jean Matsko Hood, combines complete information on the best ways to prepare coconut meat, milk, and oil. Recipes range from soups and salads to appetizers, entrées, beverages, and desserts. Hood tops it all off with a dash of poetry to present a creative cookbook that is most satisfying.

Coconut Delights Cookbook has INSTANTLY become one of my favorites!"…

Kimberly Carter
Publicist

Praise for Coconut Delights

A Collection of Coconut Recipes
Cookbook Delights Series – Book 4

"***Coconut Delights Cookbook*** has fascinating tidbits that will delight and entertain your family. With witty and factual information about coconut, as well as more than 260 wonderful tasty recipes, this cookbook promises to delight the palate and tantalize the taste buds. Also included is an enormous variety of scrumptious coconut recipes that promise to deliver a unique experience.

The history and poetry that is included will intrigue and enchant the senses while enjoying the prose."…

Mary Scripture-Smith

Graphic Designer

"Tropical cultures have relied on the coconut as a source of food, material for clothing and housing, and as a source of drink for many, many years.

Now ***Coconut Delights Cookbook*** can do the same for your kitchen. You will truly be inspired by the wealth of information, poetry, and more than 270 recipes that you will find in this cookbook.

Get this book and have fun inspiring others with your new found delicacies."…

Ed Archambeault
Spokane, WA

Coconut Delights

A Collection of Coconut Recipes
Cookbook Delights Series – Book 4

Karen Jean Matsko Hood

Coconut Delights

A Collection of Coconut Recipes
Cookbook Delights Series – Book 4

Karen Jean Matsko Hood

Published by:

Whispering Pine Press International, Inc.
Your Northwest Book Publishing Company

P.O. Box 214
Spokane Valley, WA 99037-0214 USA
Phone: (509) 928-8700 | Fax: (509) 922-9949
Email: sales@whisperingpinepress.com
Websites: www.WhisperingPinePress.com
www.WhisperingPinePressBookstore.com
Blog: www.WhisperingPinePressBlog.com
SAN 253-200X
Printed in the U.S.A.

Published by Whispering Pine Press International, Inc.
P.O. Box 214
Spokane Valley, Washington 99037-0214 USA

For sales outside the United States, please contact the Whispering Pine Press International, Inc., International Sales Department.

Manufactured in the United States of America. This paper is acid-free and 100% chlorine free.

Book and Cover Design by Artistic Design Service, Inc.
P.O. Box 1782
Spokane Valley, WA 99037-1782 USA
www.ArtisticDesignService.com

Library of Congress Number (LCCN): 2014901441

Hood, Karen Jean Matsko
Title: Coconut Delights Cookbook: A Collection of Coconut Recipes: Cookbook Delights Series – Book 4

p. cm.

ISBN: 978-1-59434-293-6 case bound
ISBN: 978-1-59434-294-3 perfect bound
ISBN: 978-1-59434-219-6 spiral bound
ISBN: 978-1-59434-295-0 comb bound
ISBN: 978-1-59434-297-4 E-PDF
ISBN: 978-1-59210-517-5 E-PUB
ISBN: 978-1-59434-806-8 E-PRC

First Edition : February 2014
1. Cookery (Coconut Delights Cookbook, A Collection of Coconut Recipes, Cookbook Delights Series – Book 4) 1. Title

Coconut Delights Cookbook

A Collection of Coconut Recipes
Cookbook Delights Series – Book 4

Gift Inscription

To:____

From: ______________________________________

Date: ______________________________________

Special Message: ______________________________

____

____

____

____

____

____

____

____

It is always nice to receive a personal note to create a special memory.

www.CoconutDelights.com
www.WhisperingPinePress.com
www.WhisperingPinePressBookstore.com

Dedications

To my husband and best friend, Jim.

To our seventeen children: Gabriel, Brianne Kristina and her husband Moulik Vinodkumar Kothari, Marissa Kimberly and her husband Kevin Matthew Franck, Janelle Karina and her husband Paul Joseph Turcotte, Mikayla Karlene, Kyler James, Kelsey Katrina, Corbin Joel, Caleb Jerome, Keisha Kalani Hiwot, Devontay Joshua, Kianna Karielle Selam, Rosy Kiara, Mercedes Katherine, Jasmine Khalia Wengel, Cheyenne Krystal, and Annalise Kaylee Marie.

To our grandchild Nola Paige and future grandchildren.

To our foster grandchildren: Courtney, Lorenzo, and Leah.

To my brother, Stephen, and his wife, Karen.

To my husband's ten siblings: Gary, Colleen, John, Dan, Mary, Ray, Ann, Teresa, Barbara, Agnes, and their families.

In loving memory of my mom, who passed away in 2007; my dad, who passed away in 1976; and my sister, Sandy, who passed away due to multiple sclerosis in 1999.

To Sandy's three sons: Monte, Bradley, and Derek. To Monte's wife, Sarah, and their children: Liam, Alice, Charlie, and Samuel and their foster children. To Bradley's wife, Shawnda, and their children: Anton, Isaac, and Isabel.

To our foster children past and present: Krystal, Sara, Rebecca, Janice, Devontay Joshua, Mercedes Katherine, Zha'Nell, Makia, Onna, Cheyenne Krystal, Onna Marie, Nevaeh, and Zada, our future foster children, and all foster children everywhere.

To the Court Appointed Special Advocate (CASA) Volunteer Program in the judicial system which benefits abused and neglected children.

To the Literacy Campaign dedicated to promoting literacy throughout the world.

Acknowledgements

The author would like to acknowledge all those individuals who helped me during my time in writing this book. Appreciation is extended for all their support and effort they put into this project.

Deep gratitude and profound thanks are owed to my husband, Jim, for giving freely of his time and encouragement during this project. Thanks are also owed to my children Gabriel, Brianne Kristina and her husband Moulik Vinodkumar Kothari, Marissa Kimberly and her husband Kevin Matthew Franck, Janelle Karina and her husband Paul Joseph Turcotte, Mikayla Karlene, Kyler James, Kelsey Katrina, Corbin Joel, Caleb Jerome, Keisha Kalani Hiwot, Devontay Joshua, Kianna Karielle Selam, Rosy Kiara, Mercedes Katherine, Jasmine Khalia Wengel, Cheyenne Krystal, and Annalise Kaylee Marie. All of these persons inspired my writing.

Thanks are due to Sharron Thompson for her assistance in editing and typing this manuscript for publication. Thanks go to Artistic Design Service, Inc. for their assistance in formatting and providing a graphic design of this manuscript for publication. This project could not have been completed without them.

Many thanks are due to members of my family, all of whom were extremely supportive during the time it took to complete this project. Their patience and support are greatly appreciated.

Coconut Delights Cookbook

Table of Contents

Coconut Delights Cookbook

A Collection of Coconut Recipes
Cookbook Delights Series – Book 4

Introduction

Coconut has always been one of my favorite fruits. They are full of flavor and texture and are a popular fruit in all forms of preparation. They are great for cooking and nutritious to eat alone

As a poet, I found it enjoyable to color this cookbook with poetry so that readers could savor the metaphorical richness of the coconut as well as its literal flavor. Also included are some articles on history, cultivation, and botanical information, along with interesting facts about coconuts. Sections that discuss health and nutrition as well as some coconut folklore are also included in this book.

The *Cookbook Delights Series* would not be complete without *Coconut Delights Cookbook* since coconuts are such a prized American fruit. We hope you enjoy reading this cookbook as well as trying out all of the delicious recipes that have been gathered together for your culinary adventures.

The cookbook is organized in convenient alphabetical sections to assist you in finding recipes related to the type of cooking you need: appetizers and dips; beverages; breads and rolls; breakfasts; cakes; candies; cookies; desserts; dressings, sauces, and condiments; jams, jellies, and syrups; main dishes; pies; preserving; salads; side dishes; soups; and wines and spirits.

Following is a collection of recipes gathered and modified to bring you *Coconut Delights Cookbook: A Collection of Coconut Recipes, Cookbook Delights Series* by Karen Jean Matsko Hood.

Coconut Botanical Classification

Coconut Botanical Classification

The coconut palm (*Cocos nucifera*) is a member of the palm family (also known as *Arecaceae)* and is a family of flowering plants. There are roughly 202 currently known genera with around 2600 species, most of which are restricted to tropical or subtropical climates. Of all the families of plants, the *Arecaceae* is the most easily recognizable. Most palms are distinguished by their large, compound evergreen leaves arranged at the top of an unbranched stem. However, many palms are exceptions to this statement and have many other physical characteristics. As well as being morphologically diverse, palms also inhabit nearly every type of habitat within their range, from rainforests to deserts.

Coconut palms can grow up to eighty feet tall. Their trunks are ringed with scars where old leaves have fallen, leaving the trunk smooth. The top of the trunk is crowned with a rosette of leaves, and there are male and female flowers that grow on the same plants on flowering branches. Flowers are pale yellow and are ½-inch long. The base of flowering branches are tapped for sap.

Their leaves are feather-shaped and split into lots of leaflets. Long leaves can grow up to 23 feet long and can have 250 leaflets.

The fruit produced by the palm tree are, obviously, coconuts. One tree can produce up to one hundred coconuts in a season.

Coconut Cultivation and Gardening

Coconut Cultivation and Gardening

Coconut trees are plentiful in the tropical regions and make an attractive plant. They provide food, nutrition, and shelter wherever they grow.

Coconuts require well-drained soil and need regular watering. They also require good light, a minimum temperature of 80 degrees F., and a good organic fertilizer.

Coconuts are prone to pests and diseases. Rhinoceros beetles can damage young trees, and they may be overgrown or smothered by weeds.

In nurseries, young coconuts with their husk still intact are laid sideways in a shady place and covered with coconut or banana leaves, then watered regularly until a sprout appears, usually after about six weeks.

In a pot at home, coconuts should be just set in the soil, leaving the skin slightly exposed. It is important that the light source is good. Water it regularly until a sprout emerges. Two leaves should develop. In nurseries, the sprouted coconuts are planted out so that the best plants can be selected for final transplanting, which takes place about five months later in a permanent position during the beginning of the rainy season.

Remove dead leaves as they drop and mulch around the base of the trees annually with compost, manure or seaweed. Harvesting can either be when the ripe nuts fall from tree or by climbing up and cutting fruits off.

The coconuts are oval and covered with a smooth skin that can be bright green, brilliant orange, or ivory colored.

When seeds germinate, the new shoot sprouts from one of the eyes.

The only two states in the United States that can grow coconut palms and reproduce outdoors without irrigation are Hawaii and Florida. The Florida Keys are the only safe haven from the cold that palm trees can live on the United States mainland. They may be grown in favored microclimates on the Barrier Islands near Brownsville, Texas but they are usually damaged or die by the occasional winter freeze in that area.

The farthest north a coconut palm has been known to grow outdoors is in Newport Beach, California. In order for coconut palms to survive in Southern California they need sandy soil, minimal water in the winter to prevent root rot and would benefit from root heating coils.

Coconut Facts

Coconut Facts

The coconut provides a nutritious source of meat, juice, milk, and oil that has fed and nourished populations around the world for generations and has been a staple food for many people.

The coconut palm is one of the most useful trees in the world because every part of it has some value.

Their leaves are woven, matted, twisted or plaited to make clothing, mats, baskets, and roofing. Their fruits provide food, drink, oil, medicine, containers, fiber for ropes and mats; and their wood helps build houses and boats.

The leaves are used for making fans, baskets, and thatch. The coarse fibers obtained from the husk, called coir, is made into cordage, mats, and stuffing. The hard shell and the husk are used for fuel. The fibrous center of the old trunk is also used for ropes, and the timber, known as porcupine wood, is hard and fine-grained and takes a high polish. Containers of various kinds are made from the nutshells, such as cups, ladles, and bowls, which are often highly polished and ornamentally carved.

The greatest value of the coconut lies in the oil, which is extracted from the dried kernels of the fruit. There are many health benefits in coconut, due to its fiber and nutritional content.

Large quantities of whole coconuts and shredded or desiccated coconut made from the dried meat which is known as copra, are exported for use in the making of cakes, desserts, and confectionery.

People from all cultures who live where the coconut palm grows have learned of the coconut's value as a food source and an effective staple in folk medicine.

Coconut Delights Cookbook
A Collection of Coconut Recipes
Cookbook Delights Series – Book 4

Coconut Folklore

Coconut Folklore

The folklore of the coconut is extensive because the coconut has always been an item of great value. Beliefs vary from fertility taboos to unseen magical forces.

It is believed that coconut juice, when drunk by expectant mothers, will help the fetus grow stronger and with greater vitality.

In northern India, it is called the fruit of the "Tree of Life." When a woman wanted to conceive, she would go to a priest to receive her special coconut. In Bali, women are forbidden to touch coconut palms for fear of draining the fertility of the tree into the woman. In South Asia, the palms are believed to be of the tree which grants all wishes. In Samoa, they do not pick up coconuts lying on the ground because this is believed to belong to some “magical spirits.” If you pick it up, an unseen spirit may punish you with a painful, incurable illness. Natives of New Guinea believe that when the first man died on the island, a coconut tree sprouted from his head.

The three eyes of the coconut represent the three eyes of the great god Shiva. Fishing communities along the coasts of peninsular India believe in appeasing the seas with offerings of coconut.

Weddings of the Nair community in Kerala are solemnized in front of a spray of coconut flowers planted in a bowl of rice.

Coconut Delights Cookbook
A Collection of Coconut Recipes
Cookbook Delights Series – Book 4

Coconut History

Coconut History

Around the year 1555, the English name “coconut” was derived from the Spanish and Portuguese word coco, which means "monkey face." Spanish and Portuguese explorers found a resemblance to a monkey's face in the three round indented markings or eyes found at the base of the coconut.

Native to the Pacific, coconuts spread to the tropics on ocean currents before humans intervened in their culture. Coconuts have been used throughout history in South Asia for a wide range of purposes.

The coconut is the most familiar palm of the tropics, yet until recently its origin was hotly disputed. It is now thought to have come from the western Pacific and spread via human activity and ocean currents to most of the tropics. Today it is a domesticated plant and has become an essential resource for food, shelter, fuel, and tools. South Asia is no exception, and coconuts are used in everyday life, particularly in south India.

Coconut palms are featured in Indian writings dating back more than two thousand years.

Today, India is the third largest producer of coconuts in the world. Coconut plays an important role in Indian rituals and mythology, for it resembles a human head with three marks on its shell like eyes and a mouth, and fiber like hair. It was known as the fruit of the gods, and cutting the tree was prohibited.

Coconut Delights Cookbook
A Collection of Coconut Recipes
Cookbook Delights Series – Book 4

Coconut Nutrition and Health

Coconut Nutrition and Health

One of the most nutritious fruits is the coconut. There are many ways to enjoy fresh coconuts. Young coconuts are considered highly nutritious and have a green shell, while mature coconuts are hairy and brown in color. Young coconuts have more water and a soft, gel-like meat, and mature coconuts have less water and firmer meat.

Coconut oil, also known as coconut butter, is the fatty oil that comes from the coconut meat. A high-quality coconut oil is the healthiest oil that you can consume.

Coconut can be safely added to most everyone's diet. Coconuts can add flavor, variety and healthy nutrients to your diet. Coconuts are rich in lauric acid, which is known for being antiviral, antibacterial and antifungal, and boosts the immune system.

Fresh coconut juice is one of the highest sources of electrolytes known to man, and can be used to prevent dehydration, instead of a sports drink. On hot days in the tropics when you feel sluggish and overheated, drink lots of young coconut juice. It will revive you and replenish your energies.

Although coconut oil is about eighty-seven percent saturated fat, in its unrefined, virgin state, it is actually beneficial. Fifty percent of its fat content consists of lauric acid, a healthy fatty acid which is critical in your well being and your ability to fight off disease.

Some of the vitamins and minerals found in fresh coconut are B1, B2, niacin, folic acid, Vitamin C, calcium, iron, magnesium, potassium, and zinc. Coconuts and coconut oil are healthy and can promote an increased metabolism, healthy thyroid function, boost your daily energy, and rejuvenate your skin and prevent wrinkles as coconut oil is a great moisturizer. Not only is it a good thirst quencher, the juice is also good for reducing heat in the body.

In folk medicine, the fresh juice of young coconuts is recommended for reducing fevers, relieving headaches, stomach upsets, diarrhea and dysentery, for strengthening the heart, and for restoring energy to weakened bodies recovering from illness.

Coconut Delights Cookbook
A Collection of Coconut Recipes
Cookbook Delights Series – Book 4

Poetry

A Collection of Poetry with Coconut Themes

Table of Contents

Page

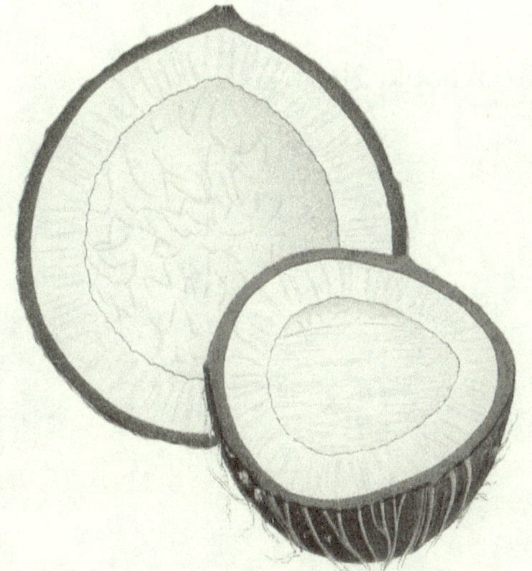

Coconuts in the Sand

Think back with me to long ago
when we were young
and time seemed slow.
The taste of salty air
lingered on my tongue.
We walked hand in hand,
our footprints solitary in the sand.
We came upon a grove of palms,
a sweet oasis with ground untouched.
Under the shade of those rustling palms
we found a few scattered coconuts.
Eagerly, we gathered the sweet fruit
and cracked them open to enjoy
the nectar that seemed sent from above.
We drank to our youth and health
and toasted our love.
The mingled taste of salt and sweet coconut
reminds me of the beginning
of our long-standing love. A love mature
that grows throughout the years.
Return with me to yesteryear;
Let's drink to life and love once more.

Karen Jean Matsko Hood ©2014
Published in Coconut Delights Cookbook, 2014
By Whispering Pine Press, Inc., 2014

Coconuts, Coconuts, Coconuts

Some lands do thrive
on this pure fruit.
The gift of the palm
is a gift in your palm.
The coconut
is great for lots of things.
They make great costumes
for island hula dancers,
or a bowl from which to drink.
Coconuts even make great sound effects
for the wayfaring traveler
without a horse.
I like to eat the sweet meat inside;
it fills my belly and makes me satisfied.
Oh how I crave it.
Coconuts, coconuts, coconuts, Oh My!

Karen Jean Matsko Hood ©2014
Published in Coconut Delights Cookbook, 2014
By Whispering Pine Press, Inc., 2014

Tough Exterior

It's hard to crack.
I can't get through.
Why won't you let me in?
I knock and knock
and still the walls stand firm.
Don't shut me out,
for if you do
you'll hold in
the hate and fear
that rots your soul
and brands your spirit
with weary service,
lets the walls crumble
before you're too far gone.
I know there's sweet tenderness
behind that tough exterior
that waits to release,
if only a crack will open.

Published in Coconut Delights Cookbook, 2014
By Whispering Pine Press, Inc., 2014

Fresh Air

I took a breath of goodness
And wanted to inhale more.
Evil entered the breeze of
My space and threatened
The air that I breathe.

Livid skids evolved.
Storm clouds and lightning
Bolts illuminated the sky.
Raindrops cleansed the rotund darkness,
So I could smell the light once more.

Published in Coconut Delights Cookbook, 2014
By Whispering Pine Press, Inc., 2014

Quiet Sand

Seagulls fly over sun bleached sand,
beached pieces of stone grountbgether over millions of years.
Single buds from the flock turn for their voice to be heard;
some whisper quiet, others speak loud. All have
a story to be told. Each waitsto be heard, to be noticed.
Instead plastic thongs march over granules that
carry disinterested humans,
blind to messages arranged below their feet.
Nature resonates mixed with wind to
keep the wings afloat of
seagulls above. One single bird
looks down with yellow eyes ever-observant.
He notices his wing is clipped
and down he falls before he realizes the message.

Published in Coconut Delights Cookbook, 2014
By Whispering Pine Press, Inc., 2014

Ocean Waves

Tropical ocean waves crash and
Fall onto sunning beached shells.
Frothy crests brew in strength
Till they force their way to boulders
Framing the seaside shores.

Salty smells mixed with
Malodorous fish punctuate the
Stinging breeze which tickles my
Sunlit face.
Hermit crabs crawl by carrying

Their houses far-too-large,
Conducting their work of the
Day, while a stray baby eel
Struggles to find its path through
The maze of rocks aligned on the shore.

Scuba divers line up
To pay their dues, exploring
Unseen ocean floors, modeling
Their newest gear. A barefoot
Native woman walks

The burning beach
With her net, to find
The caught flesh of the day.
Everyone goes about their business
Yet, they do not see the sand.

Karen Jean Matsko Hood ©2014
Published in Coconut Delights Cookbook, 2014
By Whispering Pine Press, Inc., 2014

Salt Breeze

A young girl came to look at the rolling ocean
With its seashells on the shore,
As she walked on hot sand with
Calloused and aching feet,

Isolated and melancholy she sat
Washing her tired soles.
On cool soothing waters of the ocean
Pounding against rocks.

So very strong and bold
Salt breeze stings her parched lips
Naked to the drying wind.
Ground up seashells thousands of

Years old now broken and
Altered, become grains of sand under her feet
Only to be cooled by refreshing tides.
Northwest winds provoke some laughter

Of children down the seashore,
Making sand sculptures,
Playing together, and talking to the ocean
Above the singing sea breeze.

Seashells once had a life and a home their own,
Now grains of sand
Specs of burning dust below
Tired feet. A grain of infinity

In the hourglass of time.

Published in Coconut Delights Cookbook, 2014
By Whispering Pine Press, Inc., 2014

Sea Dance

Sea creatures swim and dance
below the water, oblivious to what transpires
overhead. Wind
sandblasts the shores as
tumbleweeds carry the tiny
chameleon caught in the storm.
So much life in the ocean. Adventure
and excitement hides offshore
in miles unknown. Humans
sit in cars in traffic
jams. Ghost crabs scurry
over grains of sand to form
a banquet for the lizard.
Groupers feed in frenzies
under high tide and
feast till full. Small aquatic
animals hide while
sand plovers rest.
The driver honks his horn
in the traffic jam
and no one seems to
notice in the endless
dance of life.

Published in Coconut Delights Cookbook, 2014
By Whispering Pine Press, Inc., 2014

Tranquil Shores

Heavy winds moved in from sullen shores
While quiet waters anticipated hurricane waves
Brought from distant shores silenced by thunder.

Local villagers scurried for safety as the smell of stormy
Burned their tired nostrils. Children playing
on dirt roads ignored warnings of the clouds

While frantic mother's searched for cries
All too quiet in the noise while the waves
Crashed the boulders misshaped by centuries of past
memories.

Can there be a reason for this turn
from tranquil peace to violence and nature?
Can the mother save her children from the storm?

Or do we only fear the death
That now can pass by the quiet shores.

Published in Coconut Delights Cookbook, 2014
By Whispering Pine Press, Inc., 2014

Coconut Types

Coconut Types

There are two types of coconut trees. Three-year coconut trees are trees that bear fruit in approximately three years. The six-year coconut tree produces fruit in about six years. You can easily distinguish between the two, as the three-year coconut tree does not have a bulb at the base of the trunk, which is characteristic of the six-year coconut tree. The three year coconut trees' leaves are more uniform in shape, being as slender at the base of the leaf, up to about four to five inches from the tapering tip. The six-year coconut tree has a larger leaf at the bottom and tapers slightly through the leaf up until the tip.

There are two colors of coconuts. There is the green variety and then there is the red/orange type. On Guam there are about seven different types of coconut trees.

Coconut milk and cream is made by pouring boiling water over freshly grated meat and squeezing out the liquid. It can be diluted with water to create different thicknesses for sweet and savoury dishes and baked products. Coconut milk and cream is sold in powdered and canned forms.

Desiccated coconut is washed, steamed, shredded and dried meat that is used in sweets, baking, savoury dishes and as snack food.

Coconut oil is used for cooking and to make margarine, ice creams and sweets. Oil can be processed using fresh coconut.

Coconut water from the seed cavity is sweet, and is commercially extracted and preserved as a drink.

As with many other palms, the heart is a delicacy. It is the tender, young apex at the top of the stem, also known as palm cabbage. A sweet sap is tapped from unopened flowering branches and can be boiled to give a rich palm sugar, known as jaggery or gur.

Coconut Delights Cookbook
A Collection of Coconut Recipes
Cookbook Delights Series – Book 4

RECIPES

Coconut Delights Cookbook
A Collection of Coconut Recipes
Cookbook Delights Series – Book 4

Appetizers and Dips

Table of Contents

Page

Chicken and Coconut Shumai

Shumai are open-topped dumplings stuffed with ground meat, then steamed. This recipe is an appealing mix of sweet, hot, salty, and sour. They make popular appetizers that will disappear quickly.

Ingredients:

¼ c. unsweetened coconut milk
¼ c. carrot, coarsely shredded
½ tsp. fresh ginger, minced
½ tsp. salt
¼ tsp. ground pepper
2 Tbs. basil, chopped
2 Tbs. Asian fish sauce
2 Tbs. sugar
2 tsp. fresh lime juice
1 garlic clove, minced
2 Thai chilies, minced
1 lb. ground chicken
1 lg. egg, beaten
1 small shallot, minced
40 wonton wrappers
green leaf lettuce leaves, for steaming
chili sauce, for serving

Directions:

1. In large bowl, combine chicken with coconut milk, carrots, chilies, basil, fish sauce, sugar, lime juice, garlic, egg, shallot, ginger, and salt, and pepper.
2. Using your hands, mix thoroughly.
3. Hold a wonton wrapper in the palm of your hand; keep the rest covered with plastic wrap.
4. Place a rounded tablespoon of filling in the center of wrapper; pinch edges all around to form a cup that is open 1 inch at the top.

5. Repeat with remaining wrappers and filling.
6. Fill a wok or large skillet with 2 inches of water and bring to boil.
7. Line a double-tiered bamboo steamer with lettuce leaves and arrange the wonton in steamer without crowding.
8. Cover; steam over moderate heat until cooked through, 10 minutes.
9. Repeat with the remaining wonton wrappers.
10. Serve immediately with chili sauce.

Coconut Dumplings

These dumplings are delicious for breakfast or to serve with your favorite tea.

Ingredients:

8 Tbs. unsweetened coconut, shredded
2 Tbs. sugar
1 tsp. whole cumin seeds
1 tsp. water
2 c. rice flour
1½ c. water

Directions:

1. In medium bowl, combine first 4 ingredients.
2. In small bowl, combine flour and water.
3. Form into ball.
4. Take a ping-pong size ball of dough and flatten.
5. Place a marble-size amount of coconut mixture in middle of each piece.
6. Close dough around coconut to form a ball.
7. Roll in hand until well formed.
8. Repeat until all dough used.
9. Steam 10 minutes.

Dip for Fresh Vegetables

This dip is flavorful and great with vegetables.

Ingredients:

1 tsp. soy sauce
¼ c. sweet chili sauce
½ c. peanut butter, extra crunchy
¼ c. coconut milk powder
½ c. water, boiling

Directions:

1. In blender, add all ingredients.
2. Blend until smooth.
3. Refrigerate 2 hours.
4. Place dip in serving bowl on large platter.
5. Surround with cold, crisp vegetables.

Coconut Dip

This is an easy-to-make dip that is especially good with fresh fruit.

Ingredients:

⅔ c. sour cream
⅓ c. coconut cream
⅓ c. brown sugar, firmly packed

Directions:

1. In small bowl, mix ingredients together.
2. Use as a dip for fruits such as orange sections, fresh pineapple spears, fresh figs, apricots, or banana slices.

Crispy Coconut Shrimp

This is an easy-to-make shrimp dish that can be served as a main dish or an appetizer.

Ingredients:

½ c. flour
1 Tbs. sugar
1 tsp. ground red pepper
½ tsp. salt
2 eggs
2 Tbs. water
2½ c. coconut, sweetened, flaked
1 lb. lg. shrimp, tails left on
2 c. canola oil

Directions:

1. In medium bowl, combine flour, sugar, red pepper, and salt.
2. In small bowl, beat together eggs and water.
3. Place coconut in shallow dish.
4. Coat shrimp with flour, dip in egg.
5. Roll in coconut pressing firmly to coat both sides of shrimp.
6. Heat oil.
7. Cook shrimp in batches until golden brown.
8. Drain on a paper towel.
9. Note: A good dipping sauce is made with plain yogurt and pina colada mixer.

Did You Know? . . . .

Did you know that coconut shells can be used to make gas masks?

Coconut Beer Shrimp with Sweet and Tangy Sauce

The beer makes a nice coating for the coconut shrimp.

Ingredients for sauce:

2 c. orange marmalade
¼ c. Creole or Dijon mustard
3 Tbs. shredded horseradish

Ingredients for seasoning:

2½ Tbs. paprika
2 Tbs. salt
2 Tbs. garlic powder
1 Tbs. black pepper
1 Tbs. onion powder
1 Tbs. cayenne pepper
1 Tbs. dried leaf oregano
1 Tbs. dried thyme

Ingredients for shrimp:

4 eggs
1 c. beer
3½ tsp. seasoning, divided
1¼ c. flour
2 Tbs. baking powder
48 lg. raw shrimp, peeled, tails on, deveined
2 c. coconut, shredded
canola oil, for deep-frying

Directions for sauce:

1. In small bowl, combine marmalade, mustard, and horseradish.
2. Serve with beer shrimp.

Directions for seasoning:

1. In small bowl, combine all ingredients thoroughly.
2. Store in airtight jar or container.

Directions for shrimp:

1. In large bowl, combine eggs, 1 teaspoon Creole seasoning, flour, and baking powder.
2. Coat shrimp in beer batter and roll in coconut.
3. Fry in oil heated to 375 degrees F. until golden brown.
4. Serve hot with sauce.

Yields: 6 servings.

Dip or Dressing Base

Coconut based products add flavor to dips and dressings.

Ingredients:

2 c. coconut sour cream or sour cream (16 oz.)
1 c. coconut cream (8 oz.)
½ c. coconut vinegar, enough for right consistency
sea salt, to taste

Directions:

1. In small bowl, combine all ingredients.
2. Variations: Roasted tomato and basil with some freshly cracked pepper or cayenne for a little zip.
3. Add dill, onion, and garlic for a nice dill dip.
4. Note: For dip add ½ cup vinegar.
5. For salad dressing add ¾ cup vinegar.

Coconut Chicken Pieces

This is a delicious and easy-to-make chicken recipe to enjoy.

Ingredients:

- 1 lb. boneless chicken breast halves (4 pieces)
- 1 Tbs. cornstarch.
- 1 tsp. ground coriander
- ½ tsp. salt
- ¼ c. flour
- 2 egg whites
- 2 Tbs. honey
- 1½ c. coconut, sweetened, shredded
- 1 jar orange marmalade (12 oz.)

Directions:

1. Preheat oven to 400 degrees F.
2. Cover a baking sheet with aluminum foil; spray with vegetable oil.
3. Rinse chicken pieces under cold water.
4. Dry chicken with paper towels.
5. Cut up each chicken breast into 8 pieces, each about the size of a quarter.
6. In a plastic bag, mix together cornstarch, coriander, salt, and flour.
7. In small bowl, mix egg whites and honey.
8. Place coconut in another bowl.
9. Place chicken pieces in bag and shake.
10. Coat chicken pieces in egg and honey mixture.
11. Roll chicken in coconut.
12. Place on prepared baking sheet.
13. Bake for 10 minutes, rotate baking sheet.
14. Bake another 5 to 7 minutes, or until thermometer reads 170 degrees F.
15. Remove from oven, cool slightly.

16. Lightly cover them with orange marmalade.
17. To serve, spear the chicken balls with long toothpicks.

Coconut Fruit Kabobs with Coconut Dip

Make them ahead for picnics, barbeques, and parties. They are healthy, and everyone seems to enjoy them. Make more than you think you'll need!

Ingredients for dip:

1½ c. vanilla yogurt
3 Tbs. coconut, flaked
1½ Tbs. orange marmalade

Ingredients for kabobs:

1 med. red apple, unpeeled
1 med. pear, unpeeled
1 Tbs. lemon juice
20 unsweetened pineapple chunks
20 seedless red or green grapes
20 fresh strawberries, capped
20 wooden skewers

Directions for dip:

1. In small bowl, combine all ingredients and stir well.
2. Store in refrigerator; serve with the kabobs.

Directions for kabobs:

1. Cut apple and pear into 20 bite-size pieces.
2. In a bowl, place fruit and toss with lemon juice gently.
3. On a skewer, alternate an apple, pear, grape, pineapple, and a strawberry.

Chicken Satay with Peanut Sauce

Chicken satay is one of our favorite family appetizers. Enjoy.

Ingredients for chicken:

½ lb. chicken fillet
2 Tbs. olive oil
½ c. coconut milk
2 Tbs. coriander roots
½ tsp. curry powder
2 tsp. sugar
½ Tbs. fish sauce

Ingredients for sauce:

3 Tbs. olive oil
2 Tbs. red curry paste
½ tsp. ground coriander
2 c. coconut milk
1 c. coarsely ground peanuts
2 Tbs. sugar
2 Tbs. lime juice
1 tsp. salt
1 Tbs. peanut butter

Directions for chicken:

1. Slice chicken into thin strips.
2. In medium bowl, mix together all other ingredients.
3. Add chicken slices.
4. Marinate for several hours.
5. Put chicken on metal or wooden skewers.
6. Grill on fire or electric grill turning the skewers regularly.

Coconut Cream

This coconut cream is a pleasant contrast for the acidic pineapple.

Ingredients:

1 can pineapple chunks, drained, reserve juice (20 oz.)
4 tsp. cornstarch
⅔ c. coconut cream
1 tsp. vanilla extract
1 c. whipping cream, whipped (½ pint)

Directions:

1. In medium saucepan, combine reserved pineapple juice and cornstarch.
2. Stir in coconut cream.
3. Over medium heat, cook and stir until slightly thickened.
4. Stir in vanilla.
5. Cool to room temperature.
6. Fold in whipped cream.
7. Chill.
8. Serve with pineapple chunks and other fruit as an appetizer, dessert, or salad.
9. Refrigerate leftovers.

Directions for microwave:

1. In 1-quart glass measure with handle, dissolve cornstarch in pineapple juice.
2. Stir in coconut cream.
3. Heat on 100% power (high) 3 to 4 minutes, or until thick and bubbly, stirring twice.
4. Proceed as above.

Coconut Pork Satay

Our family loves satay as an appetizer, and this is an excellent pork version. This is delicious served with peanut dipping sauce (recipe below).

Ingredients for marinade:

1 lb. lean boneless pork loin
½ c. coconut milk
2 Tbs. fresh cilantro, chopped
1 tsp. turmeric
½ tsp. salt
8 bamboo skewers

Ingredients for sauce:

½ c. unsalted roasted peanuts
1 Tbs. peanut oil
1 ½-inch-thick slice fresh ginger
⅓ c. canned unsweetened coconut milk
2 tsp. soy sauce
4 tsp. fish sauce
1 tsp. sugar
1 Tbs. fresh lime juice
½ c. finely minced cilantro leaves and stems
2 fresh chilies
4 garlic cloves
pinch of salt

Directions for marinade:

1. Cut pork into eight 1 x 4-inch strips.
2. In medium bowl, combine milk, cilantro, turmeric, and salt.
3. Add pork strips.
4. Cover and refrigerate 1 hour.
5. Preheat grill.

6. Soak bamboo skewers in cold water.
7. Drain pork and thread on skewers.
8. Grill until thoroughly cooked.
9. Serve with Thai Peanut Dipping Sauce.

Directions for sauce:

1. In blender or food processor, add peanuts with peanut oil.
2. Blend, on high, until peanuts form a rough paste.
3. Add rest of ingredients except for cilantro.
4. Blend until smooth.
5. Stir in cilantro.
6. Thin with peanut oil to taste.

Sweet and Spicy Dipping Sauce

If you enjoy hot, spicy, and sweet you will enjoy this sauce.

Ingredients:

⅓ c. coconut cream
⅓ c. chili sauce
1 Tbs. lemon juice concentrate
2 tsp. soy sauce
1 tsp. Worcestershire sauce
1 tsp. prepared horseradish
½ tsp. hot pepper sauce

Directions:

1. In small bowl, combine all ingredients; mix well.
2. Cover; refrigerate 4 hours or overnight.
3. Serve chilled or at room temperature with chilled cooked shrimp, crab cakes, chicken fingers, and/or coconut shrimp.
4. Refrigerate leftovers.

Coconut Shrimp with Tamarind Ginger Sauce

This tangy tamarind sauce is a nice contrast to coconut shrimp.

Ingredients for sauce:

1½ Tbs. fresh lime juice
⅔ c. mayonnaise
1½ Tbs. honey
¼ tsp. salt
2 tsp. Dijon mustard
1 tsp. fresh ginger, peeled, finely grated
1 tsp. tamarind concentrate

Ingredients for shrimp:

¾ c. beer, not dark
¾ tsp. baking soda
½ tsp. salt
1 tsp. cayenne pepper
1 lg. egg
6 c. canola oil
4 c. coconut, sweetened, flaked (10 oz.)
1 c. flour
48 med. shrimp (1½ lb.) peeled, leaving tail and first segment of shell intact, and, if desired, deveined

Directions:

1. In small bowl, whisk tamarind concentrate into lime juice until dissolved.
2. Stir in remaining sauce ingredients; cover.
3. Chill.
4. Prepare shrimp.
5. Coarsely chop coconut and transfer half to a shallow soup bowl or pie plate.

6. In small bowl, whisk together flour, beer, baking soda, salt, pepper, and egg.
7. Heat oil in a 4 to 6-quart deep, heavy pot over moderately high heat until it registers 350 degrees F. on deep-fat thermometer.
8. While oil is heating, coat shrimp.
9. Hold one shrimp by tail and dip into batter, letting excess drip off, then dredge in coconut, coating completely and pressing gently to adhere.
10. Transfer to a plate; coat remaining shrimp adding remaining coconut to bowl as needed.
11. Fry shrimp in batches of 8, turning once, until golden, about 1 minute.
12. Transfer with a slotted spoon to paper towels to drain.
13. Season lightly with salt.
14. Skim any coconut from oil and return oil to 350 degrees F. between batches.
15. Serve shrimp with sauce.

Yields: 8 hors d'oeuvre servings.

Corn and Coconut Salsa

This salsa is a refreshing change from the usual salsa.

Ingredients:

2 c. corn
1 c. coconut, shredded
½ c. onion, chopped
¼ c. fresh cilantro, chopped

Directions:

1. In small bowl, toss all ingredients together.
2. Refrigerate at least 1 hour.

Coconut Seafood Stuffed Shiitake Mushrooms

These stuffed mushrooms are delicious and worth the time to make.

Ingredients:

16 lg. fresh shiitake mushrooms, stems removed
1 c. heavy cream
½ c. fresh spinach, chopped
2 Tbs. shallots, chopped
½ c. coconut milk
¼ tsp. dried dill
¼ tsp. dried tarragon
1 c. crumbled Ritz crackers
¾ c. cooked crabmeat
¼ c. cooked shrimp, chopped
¼ tsp. salt
¼ tsp. pepper
3 Tbs. grated Parmesan cheese
coconut flakes

Directions:

1. Preheat oven to 350 degrees F.
2. In saucepan, over medium-low heat, cook cream and spinach together until cream is reduced by half.
3. Stir in shallots, cook 1 minute.
4. Remove from heat.
5. Stir in milk, dill, tarragon, cracker crumbs, crab, shrimp, salt, and pepper.
6. Stuff mushroom caps with seafood mixture.
7. Place in shallow baking pan.
8. Sprinkle with cheese.
9. Bake 8 to 10 minutes.

Pork Satay

This makes a great appetizer.

Ingredients:

1 Tbs. curry powder
1 tsp. turmeric
1 Tbs. brown sugar
2 Tbs. fish sauce
2 Tbs. lime juice
1 Tbs. olive oil
1 lb. boneless pork loin, 3 x 1 x ¼-inch thick strips
½ c. thick coconut cream
1 can unsweetened coconut milk
24 8-inch bamboo skewers, soaked in water 1 hour

Directions for coconut cream:

1. Pour unsweetened coconut milk into a tall glass.
2. Allow to sit for at least 1 hour so the thick cream rises to the top.
3. Skim off cream and set aside.
4. The remaining liquid is thin coconut milk.

Directions for pork:

1. Combine curry powder, turmeric, sugar, fish sauce, lime juice, and oil.
2. Toss pork strips with marinade.
3. Thread 3 or 4 pork strips onto skewers.
4. Marinate 30 minutes.
5. Preheat grill.
6. Brush strips with thick coconut cream.
7. Place brushed side down over hot coals, 1 to 2 minutes, or until charred and cooked.
8. Turn over, brush with coconut cream and grill until cooked.

Coconut Wings with Peanut Dipping Sauce

My sons love chicken wing appetizers, and these are great with the Thai dipping sauce.

Ingredients for marinade:

2 c. coconut milk
1 med. onion, coarsely chopped
2 Tbs. garlic, crushed
2 tsp. ground turmeric
2 tsp. red pepper flakes
1 tsp. ground galanga or ½ tsp. freshly grated ginger
1 Tbs. kosher salt
20 chicken wings, jointed

Ingredients for sauce:

½ c. unsalted roasted peanuts
1 Tbs. peanut oil
1 pc. fresh ginger, 1½ inch-thick slice
⅓ c. canned unsweetened coconut milk
2 tsp. fish sauce
1 tsp. sugar
1 Tbs. fresh lime juice
½ c. finely minced cilantro leaves and stems
2 fresh chilies
4 garlic cloves
pinch of salt

Directions for marinade:

1. In blender or food processor, grind all ingredients except chicken wings into a paste the consistency of thin yogurt.
2. Reserve small amount for basting.
3. Transfer remaining marinade to glass bowl.
4. Add cleaned and trimmed chicken wings.
5. Toss liberally, cover with plastic wrap.
6. Marinate in refrigerator overnight.
7. Shake excess marinade off wings and grill over medium heat, 5 minutes per side or until crispy.

8. Brush frequently with reserved marinade.
9. Be careful not to burn wing tips.
10. The wings should be a deep mahogany color with a crispy texture when done.

Directions for sauce:

1. In blender or food processor, add peanuts with peanut oil.
2. Blend on high, until peanuts form a rough paste.
3. Add rest of ingredients except for cilantro.
4. Blend until smooth; stir in cilantro.
5. Thin with peanut oil to taste.

Mustard Dipping Sauce

Coconut flavor combines with mustard for a unique taste.

Ingredients:

2 tsp. dry mustard
½ tsp. salt
½ c. coconut milk or half and half cream
½ c. coconut cream
2 egg yolks, beaten
1 Tbs. cider vinegar

Directions:

1. In small saucepan, combine mustard and salt.
2. Gradually add cream; mix until smooth.
3. Add coconut cream and egg yolks; mix well.
4. Over medium heat, cook and stir constantly until thick and bubbly.
5. Remove from heat; stir in vinegar.
6. Cool; cover.
7. Refrigerate 4 hours or overnight.
8. Serve chilled or at room temperature with your favorite appetizers.
9. Refrigerate leftovers.

Yields: 1 cup.

Coconut Shrimp

These coconut shrimp make popular appetizers. They are easy to make and can be made ahead of time.

Ingredients:

½ tsp. salt
½ c. coconut, freshly grated or shredded
1 Tbs. cornstarch
1 tsp. ground coriander
1 c. peanut or olive oil
1 lb. raw shrimp, peeled, deveined
2 eggs

Directions:

1. In food processor, coarsely grind shrimp.
2. Transfer to medium bowl.
3. Blend in eggs.
4. Add coconut, cornstarch, coriander, and salt.
5. Mix well.
6. Refrigerate 30 minutes.
7. Heat oil in wok or large, heavy skillet over medium-high heat.
8. Add shrimp mixture to oil by heaping teaspoons, do not crowd.
9. Cook until lightly browned, 3 minutes, turning occasionally.
10. Remove with slotted spoon.
11. Drain on paper towel.
12. Repeat with remaining shrimp mixture.
13. Transfer to a serving platter.
14. Serve immediately with your favorite sauce.
15. Note: You may refrigerate until serving time and heat 10 minutes in a 450 degree oven.

Beverages

Table of Contents

Coconut Bubble Pearl Tea

Some bubble tea shops use milk powder rather than fresh milk or a combination of the two to achieve the desired creaminess.

Ingredients:

¼ c. half and half cream
½ c. chilled, cooked bubble tea pearls
¾ c. coconut milk
1 c. crushed ice
2 tsp. fresh lime juice
1 c. canned or fresh litchis, mango, or papaya
honey or sugar, to taste

Directions:

1. In a large parfait glass, place pearls.
2. In blender, combine all remaining ingredients.
3. Blend, 1 minute until mixture is frothy and creamy.
4. Pour into glass.
5. Serve with extra thick straws.

Yields: 1 16 oz. drink.

Chocolate Banana Coconut Shake

This is a pleasant shake with the flavors of coconut, banana, and chocolate.

Ingredients:

3 c. milk
1 c. coconut cream
2 bananas, sliced
¼ c. chocolate-flavored syrup

Directions:

1. In blender, combine ingredients; blend until smooth.
2. Garnish as desired.
3. Serve immediately.
4. Refrigerate leftovers.

Yields: 4 servings.

Tropical Milkshake

This is a refreshing tropical milkshake. Make sure you use the freshest fruit you can find for the best flavor.

Ingredients:

2½ c. milk
1¼ c. canned unsweetened coconut milk
1 lg. pineapple, peeled, cored, chopped
5 c. vanilla ice cream
5 ripe bananas, peeled
3 med. papayas, peeled, seeded, chopped
32 strawberries, hulled

Directions:

1. In blender, add 1¼ cups ice cream, ½ cup plus 2 tablespoons milk, 5 teaspoons coconut milk, ¼ of pineapple, 1¼ bananas, ¼ of papayas, and 8 strawberries.
2. Blend until smooth.
3. Pour into large pitcher.
4. Repeat with remaining ingredients in 3 more batches.

Yields: 20 cups.

Homemade Coconut Cream

This delicious milk can be used in making curries, confections, dessert, and beverages.

Ingredients:

1 med. size coconut
liquid from coconut, plus enough milk to equal 1 qt.

Directions:

1. Open coconut and reserve liquid.
2. Grate meat moderately fine and place in a large, heavy saucepan with liquid mixture.
3. Heat slowly, stirring constantly until mixture boils.
4. Remove from heat and cover.
5. Cool to room temperature.
6. Press through a double thickness of cheesecloth or fine sieve, forcing out as much liquid as possible.
7. Discard coconut pulp.
8. Store in refrigerator.

Yields: 1 quart.

Coconut Lime and Mint Lassi

Lime and mint add wonderful flavor to this version of the delicious lassi drink.

Ingredients:

7 oz. coconut milk
1 lime, roughly chopped
2 Tbs. honey
3 oz. ice
fresh mint, to garnish

Directions:

1. In blender or food processor, place milk and lime.
2. Process until lime is finely chopped.
3. Add honey and ice.
4. Blend until smooth.
5. Pour into a tall glass.
6. Decorate with a sprig of mint and serve.

Green Coconut Drink

Serve this coconut drink right in the coconut shells. This is an excellent drink for nursing mothers and young babies.

Ingredients:

6 green coconuts
4 lemon leaves

Directions:

1. Halve coconuts.
2. Pour coconut water into bowl, saving shells.
3. Scoop flesh out of coconuts.
4. In same bowl, mix with coconut water.
5. Prepare fibers from inside of coconut palm mid rib to whip it.
6. Cut fibers very thin; they are stiff and act like a binder to cut the coconut flesh.
7. Blend water and flesh until cut into small pieces.
8. Put mixture in saucepan.
9. Bring to boil.
10. Add lemon leaves.
11. Simmer gently 15 minutes.
12. Serve hot or cold, using coconut shells as a cup.

Coconut Drink

This makes a rich tropical drink you can well enjoy.

Ingredients:

1 banana
1 inch of fresh vanilla
2 oranges
half a coconut and all its water
half a lime

Directions:

1. Juice oranges and lime.
2. In blender, add juice and all other ingredients.
3. Blend until smooth.
4. Strain or drink as is.
5. Note: If you have a juicer, juice coconut and banana.
6. Add juice to the mixture, stirring back in as much pulp as you like.

Tropical Mockalada

This is a version of a pina colada without any liquor. Adults and children will enjoy the flavor.

Ingredients:

1⅓ c. brewed tea, chilled
1⅓ c. crushed pineapple in natural juice, chilled
1⅓ lg. ripe banana
1 c. cream or half and half
12 Tbs. cream of coconut
4 tsp. lime juice
2 c. ice cubes
pineapple, for garnish

Directions:

1. In blender, combine all ingredients except ice cubes.
2. Process at high speed until blended.
3. Add ice cubes, one at a time.
4. Process, until ice cubes are blended.
5. Garnish, if desired, with pineapple.
6. Core swizzle stick or pineapple chunks threaded on a small paper parasol.

Yields: 4 servings.

Coconut Mango Lassi

Our family loves lassi's and this is a great combination of coconut and mango.

Ingredients:

2 c. natural yogurt
6 tsp. sugar
3 ice cubes
1 Tbs. lemon juice
1 c. coconut milk, cold
1 c. mango pulp
4 mint leaves

Directions:

1. In small bowl, dissolve sugar in a little hot water.
2. In a blender, crush ice cubes.
3. Add rest of ingredients except the mint.
4. Blend until smooth.
5. Pour into glasses.
6. Garnish with a mint leaf and serve.

Yields: 4 servings.

Coconut Delight

This is a delightful cool drink to enjoy on a hot summer day.

Ingredients:

1 c. fresh orange juice
2 c. papaya pieces
6 Tbs. coconut milk
2 tsp. sugar
6 ice cubes

Directions:

1. In blender, add papaya pieces, orange juice, coconut milk, sugar, and ice cubes.
2. Blend until smooth.
3. Pour into individual glasses.
4. Serve immediately with crushed ice.

Coconut Bubble Tea

This is a favorite drink with our family. Make plenty because it will disappear quickly on a hot day.

Ingredients:

¾ c. hot brewed black tea
⅓ can coconut milk
½ c. ice cubes
1 Tbs. creamer
2 Tbs. honey

Directions:

1. In cocktail shaker, place ice cubes.
2. Add tea, milk, honey, and creamer.

3. Shake until mixture is frothy.
4. Serve.

Coconut Orange Smoothie

Coconut and orange flavors make this a delicious drink.

Ingredients:

1 can coconut milk
1 c. ice cubes
2 c. fresh orange juice

Directions:

1. In blender, add all ingredients.
2. Blend until smooth.
3. Serve cold.

Coconut Peach Pina Colada

This pina colada is excellent with fresh peaches. Enjoy!

Ingredients:

3 fresh peaches
1 c. coconut milk
4 tsp. sugar
crushed ice

Directions:

1. Place peaches in boiling water for 2 minutes.
2. Remove skin and chop.
3. In blender, add peaches, coconut milk, and sugar.
4. Blend until smooth.
5. Serve chilled with crushed ice.

Coconut Summer Swirl

This is a refreshing drink on a hot day.

Ingredients:

3 oz. pineapple juice
1 oz. cream of coconut
2 oz. ice cream
1 oz. puréed frozen strawberries
crushed ice

Directions:

1. In a blender, add first 3 ingredients.
2. Blend until smooth.
3. Pour into a glass coated with strawberry purée.

Coconut Vanilla Malt

My husband and I are especially fond of malted milkshakes, and coconut adds a nice flavor.

Ingredients:

1½ c. milk
4 scoops vanilla ice cream
1 Tbs. vanilla extract
2 Tbs. malted milk powder
1 c. coconut, shredded

Directions:

1. In a blender, combine all ingredients.
2. Blend until smooth.
3. Serve immediately.

Coconut Whipped Cream

This cream is delicious in coffee, tea, or any warm drink. Try it on your favorite dessert.

Ingredients:

coconut cream, cold
vanilla extract
sugar

Directions:

1. In medium mixing bowl, add cream.
2. On high speed, whip cream until soft peaks form.
3. Add sugar and vanilla to taste.
4. Serve.

Easy Coconut Milk

This is easy to make to keep on hand for your favorite recipes. It is also good to drink as is.

Ingredients:

2½ c. dried coconut
3 c. milk

Directions:

1. Combine coconut with milk in heavy saucepan.
2. Heat slowly, bring to a simmer.
3. Remove from heat and cool.
4. Strain milk, pressing down on coconut meat to extract all liquid.
5. Squeeze coconut through a towel.
6. Discard coconut.

Fresh Coconut Tropical Drink

This is a delicious tropical drink using fresh coconut.

Ingredients:

1 banana
2 oranges
1 inch of fresh vanilla
half a coconut and all its water
half a lime

Directions:

1. Prepare coconut.
2. Pour all fluid inside coconut into a bowl.
3. Juice oranges and lime; add juice to blender.
4. Add all other ingredients; blend until smooth.
5. Strain, or drink as is.

Homemade Coconut Milk

This makes a delicious milk to use as needed.

Ingredients:

2½ oz. grated creamed coconut or grated flesh of 1 fresh coconut
1¾ c. hot water

Directions:

1. Place coconut in a bowl.
2. Add hot water.
3. Stir the mixture to blend and cool.
4. Pour through cheesecloth and squeeze to extract as much juice as possible.
5. Store in refrigerator.

Iced Coconut Chocolate and Coffee Drink

This is a rich and delicious drink. Make sure you blend until all ingredients are smooth.

Ingredients:

⅔ c. coconut, shredded
¾ c. double strength coffee
⅓ c. chocolate syrup
½ c. half and half cream
½ c. milk
1 Tbs. sugar
2 c. ice
sweetened whipped cream

Directions:

1. Preheat oven to 300 degrees F.
2. Spread shredded coconut on a baking sheet.
3. Toast coconut in oven, stirring every 10 minutes.
4. Toast until light brown, 25 to 30 minutes. Cool.
5. Make double-strength coffee by using 2 Tbs. of coffee per cup; cool.
6. In blender, combine cold coffee, milk, ⅓ c. of toasted coconut, chocolate syrup, and sugar.
7. Blend 15 to 20 seconds; add ice.
8. Blend until ice is crushed and the drink is smooth.
9. Pour into two 16 oz. glasses.
10. Garnish each drink with whipped cream, a drizzle of chocolate, and a pinch of remaining toasted coconut.

Did You Know? . . . .

Did you know that coconut leaves can be made into toys?

Orange Coconut Frost

This is a very flavorful drink on a hot day.

Ingredients:

1 can cream of coconut (15 oz.)
1 can frozen orange juice concentrate, thawed (12 oz.)
1 tsp. vanilla extract
4 c. ice cubes
mint leaves and orange slices (optional)

Directions:

1. In blender, combine cream, juice concentrate, and vanilla; blend well.
2. Gradually add ice, blending until smooth.
3. Garnish with mint and orange if desired.
4. Serve immediately.
5. Refrigerate leftovers.

Yields: 5 cups.

Did You Know? . . . .

Did you know that there is a proper way to crack a coconut? The easiest way is to pierce the softest "eye" with a skewer and drain the water into a glass. Wrap the coconut in a tea towel and using a cleaver or hammer hit the coconut with a fair amount of force. The coconut should break open, cracking in more than one place.

Did you know that coconut trees are palm trees that grow up to 100 feet high?

Coconut Delights Cookbook

A Collection of Coconut Recipes
Cookbook Delights Series – Book 4

Breads and Rolls

Table of Contents

Carrot and Spice Bran Muffins

These moist and tasty muffins are full of things that are good for you.

Ingredients:

- 2 c. whole all-bran cereal
- 1¼ c milk
- ⅓ c. olive oil
- 2 eggs, beaten
- 1½ c. carrots, finely shredded
- ½ c. coconut
- ½ c. raisins
- 1¼ c. flour
- ½ c. brown sugar, packed
- ¼ c. sugar
- 2 tsp. baking powder
- 1 tsp. baking soda
- 1½ tsp. cinnamon
- ½ tsp. salt

Directions:

1. Preheat oven to 375 degrees F.
2. Grease or paper-line muffin cups.
3. In large mixing bowl, combine cereal, milk, oil, and eggs.
4. Let stand 10 minutes.
5. Stir in carrot, coconut, and raisins.
6. In another bowl, combine flour, brown sugar, sugar, baking powder, soda, cinnamon, and salt.
7. Add cereal mixture to flour mixture.
8. Stir until just combined.
9. Fill each muffin cup ⅔ full.
10. Bake 15 to 20 minutes, until toothpick inserted comes out clean.
11. Serve warm.

Carrot Coconut Bread

Coconut, carrot, and allspice make this bread flavorful and moist.

Ingredients:

- 2½ c. flour
- 1 c. sugar
- 1 tsp. baking soda
- 1 tsp. baking powder
- 1 tsp. cinnamon
- ½ tsp. allspice
- ½ tsp. salt
- 3 eggs
- ¾ c. canola oil
- ½ c. milk
- 2 c. carrots, finely grated
- ½ c. coconut, flaked
- ½ c. raisins
- 1½ c. pecans, chopped

Directions:

1. Preheat oven to 350 degrees F.
2. Grease and flour two 8½ x 4½ x 2½-inch loaf pans.
3. In large bowl, sift sugar, baking soda, baking powder, cinnamon, allspice, and salt; set aside.
4. In medium bowl, beat eggs.
5. Add oil and milk; blend well.
6. Stir wet ingredients into flour mixture slowly.
7. When mixed, add carrot, coconut, raisins, and pecans.
8. Pour evenly into pans.
9. Bake 40 to 50 minutes, until inserted toothpick comes out clean.

Yields: 2 medium loaves.

Coconut and Jam Muffins

Try these moist muffins with a surprise of your favorite jam.

Ingredients:

2½ c. flour
3 tsp. baking powder
½ c. sugar
1½ c. coconut
1 tsp. vanilla extract
1¼ c. milk
1 egg
3 Tbs. jam, select your favorite
7 Tbs. canola oil
extra coconut, for garnish

Directions:

1. Preheat oven to 400 degrees F.
2. Prepare muffin pan.
3. In small bowl, sift flour and baking powder together.
4. Stir in sugar and coconut.
5. In large bowl, beat egg, vanilla, milk, and oil together.
6. Add dry mixture to wet mixture until just combined.
7. Place a large tablespoon of batter into each muffin cup.
8. Add a teaspoon of jam to each.
9. Add rest of batter to each cup.
10. Sprinkle tops with coconut.
11. Bake 25 to 30 minutes.
12. Turn out onto wire rack to cool.

Yields: 9 to 10 muffins.

Coconut and Pineapple Upside-Down Muffins

Pineapple and coconut make a great addition to these easy-to-make muffins.

Ingredients:

¼ c. brown sugar, firmly packed
¼ c. olive oil
1 lg. egg white
6 Tbs. pineapple preserves, heaping
½ c. milk
1 c. flour
1½ tsp. baking powder
½ tsp. cinnamon
¼ tsp. salt
1 c. coconut, sweetened, flaked, toasted lightly

Directions:

1. Preheat oven to 400 degrees F.
2. Lightly grease a muffin tin.
3. Drop 1 heaping teaspoon preserves into each muffin cup.
4. In bowl, whisk together sugar, oil, and egg white until smooth.
5. Whisk in milk.
6. In another bowl, whisk flour, baking powder, cinnamon, salt, and coconut.
7. Add milk mixture.
8. Stir until just combined.
9. Divide batter among cups.
10. Bake 20 minutes, until inserted toothpick comes out clean.
11. Cool 3 minutes.
12. Remove from pan onto wire rack.

Coconut Banana Bread

Try this moist coconut and banana bread.

Ingredients:

1½ c. coconut, flaked
¼ c. butter, softened
⅔ c. sugar
2 eggs, room temperature
3 Tbs. milk
1 tsp. lemon juice
½ tsp. almond extract
2 c. flour
1 tsp. baking powder
½ tsp. baking soda
½ tsp. salt
1 c. bananas, ripe, mashed

Directions:

1. Preheat oven to 350 degrees F.
2. Lightly grease and flour a 9 x 5-inch loaf pan.
3. Spread coconut on a baking sheet.
4. Toast in oven for about 15 minutes, until lightly browned. Watch the coconut carefully, as it burns easily. Set aside to cool.
5. In blender or food processor, process bananas into a liquid.
6. Add butter, sugar, eggs, milk, lemon juice, and almond extract; blend together.
7. In small bowl, combine flour, baking powder, baking soda, salt, and the coconut.
8. Add to liquid mixture.
9. Pulse again only to mix the ingredients.
10. Pour mixture into prepared loaf pan.
11. Bake 1 hour, until bread is lightly brown.
12. Cool in pan, 10 minutes.

13. Turn onto wire rack.
14. Cool completely before serving.

Coconut Biscuits

These biscuits are a surprise with the great combination of papaya and coconut.

Ingredients:

- 2¼ c. flour
- 1 Tbs. baking powder
- ½ tsp. baking soda
- ½ tsp. salt
- 6 Tbs. butter, cut into pieces
- 1¼ c. coconut milk
- 1 c. fresh coconut, sweetened, grated, toasted
- ⅔ c. dried papaya, chopped
- 3 Tbs. butter, melted

Directions:

1. Preheat oven to 450 degrees F.
2. In large bowl, sift dry ingredients.
3. Add butter and blend gently.
4. Remove 3 tablespoons coconut for topping.
5. Add milk, coconut, and papaya.
6. Stir together, adding additional milk if necessary.
7. Turn out onto lightly floured surface.
8. Knead lightly for 30 seconds.
9. Roll dough ¾-inch thick.
10. With a cutter, punch out 2-inch rounds.
11. Place on lightly buttered sheet.
12. Brush tops of biscuits with butter.
13. Top with reserved coconut.
14. Bake 12 to 15 minutes.

Coconut Bread

This bread is best served hot out of the oven.

Ingredients:

1½ c. milk
½ tsp. vanilla extract
½ tsp. coconut extract
1 c. sugar
3 c. flour
1 Tbs. baking powder
1 Tbs. butter
1½ c. coconut, flaked or shredded
1 egg, beaten to blend
pinch of salt

Directions:

1. Preheat oven to 350 degrees F.
2. Lightly grease a 9 x 5-inch loaf pan.
3. In mixing bowl, add dry ingredients and stir together.
4. Add egg, milk, and extracts.
5. Mix until just blended.
6. In shallow baking pan or baking sheet, smear with the tablespoon of butter and sprinkle with coconut.
7. Toast lightly in moderate oven until coconut is golden, stirring once or twice.
8. Turn batter into greased loaf pan.
9. Bake 45 to 60 minutes, until inserted toothpick comes out clean.
10. Loosen edges of loaf; invert onto wire rack to cool.
11. Note: This will keep very well in the refrigerator for a week or two and it also freezes well.

Yields: 1 loaf.

Coconut Bread with Coconut Butter

This coconut nut bread is great served warm with the coconut butter.

Ingredients for bread:

2 c. flour
1¾ tsp. baking powder
¼ tsp. salt
1⅓ c. sugar
¼ tsp. baking soda
⅔ c. canola oil
⅔ c. buttermilk
1 c. coconut, flaked
2 lg. eggs
1½ tsp. coconut extract
1 c. pecans, chopped

Ingredients for butter:

1 c. butter
¼ c. cream of coconut

Directions for bread:

1. Preheat oven to 325 degrees F.
2. Lightly spray with oil two 9 x 5 x 3-inch loaf pans.
3. Mix all dry ingredients together.
4. Add remaining ingredients.
5. With electric mixer, mix well.
6. Bake 1 hour.
7. Serve warm with coconut butter below.

Directions for butter:

1. In small bowl, add ingredients.
2. Mix with electric mixer until fluffy.
3. Serve with coconut bread.

Coconut and Ginger Bread

Ginger and coconut flavor this bread to be made in a bread machine.

Ingredients:

3 c. bread flour
1¼ c. milk
3 Tbs. butter, unsalted
2 Tbs. honey
2 tsp. active dry yeast
1½ Tbs. fresh ginger, chopped
1 tsp. salt
¾ c. coconut, flaked

Directions:

1. In small saucepan, scald milk; remove from heat.
2. Stir in butter and honey.
3. Cool to room temperature.
4. Add milk mixture and all remaining ingredients, except coconut, in the order suggested by your bread machine manual.
5. Process on basic bread cycle.
6. At the beeper, before the end of the first kneading cycle, add coconut.
7. Let bread cool before slicing.

Coconut Cinnamon Bread

This is a very flavorful cinnamon and coconut bread.

Ingredients:

2 eggs
1¼ c. milk
1½ tsp. vanilla extract
2½ c. flour
2 tsp. baking powder
½ tsp. salt
2½ tsp. cinnamon
1¼ c. sugar

2½ c. coconut, shredded
⅓ c. butter, melted

Directions:

1. Preheat oven to 350 degrees F.
2. Grease and flour a loaf pan.
3. In small bowl, whisk together eggs, milk, and vanilla.
4. In large bowl, sift together flour, baking powder, salt, and cinnamon.
5. Stir in sugar and coconut.
6. Make a well in center, pour in egg mixture.
7. Stir until just combined.
8. Add butter, stir until just smooth, being careful not to over mix.
9. Bake 1 hour, until tester comes out clean.
10. Cool 10 minutes in pan; transfer to wire rack.

Coconut English Muffin Bread

This is an easy recipe to make in your bread machine. Enjoy!

Ingredients:

1¼ c. water
3 c. bread flour
2 tsp. sugar
3 Tbs. nonfat dry milk
1 tsp. salt
2 tsp. yeast
¼ tsp. baking soda
1 c. coconut

Directions:

1. Put all ingredients in machine in order suggested by your bread machines manufacturer, using regular bread cycle.
2. Note: This recipe makes a large loaf (1½ lb.); adjust ingredients for a medium or small size loaf.
3. Top of loaf should come out sunken; this is normal.

Coconut Pineapple Dinner Rolls

These rolls are delicious. Serve with your favorite dish or alone.

Ingredients:

- ¼ c. pineapple juice, warm
- ¼ c. water
- ¼ c. butter, cubed
- ¼ c. nonfat dry milk powder
- 1½ tsp. salt
- 3¼ c. flour
- 2¼ tsp. active dry yeast
- ¾ c. coconut, flaked
- 1 can pineapple, crushed, undrained (8 oz.)
- 1 egg
- 1 tsp. sugar

Directions:

1. In bread machine, add all ingredients except coconut.
2. Use dough setting.
3. Just before final kneading or at the beep, add coconut.
4. When dough is ready, turn onto lightly floured surface.
5. Cover with plastic wrap.
6. Let rest 10 minutes.
7. Divide into 15 portions; roll each into a ball.
8. Place in greased 9 x 13-inch baking pan.
9. Cover; let rise in a warm place for 45 minutes or until doubled.
10. Preheat oven to 375 degrees F.
11. Bake 15 to 20 minutes, until golden brown.

Yields: 15 rolls.

Lemon Coconut Bread

This lemon glazed coconut is a refreshing combination of flavors.

Ingredients for bread:

½ c. butter, softened
1 c. sugar
5 tsp. lemon juice
2 eggs
1½ c. flour
1 tsp. baking powder
¼ tsp. salt
½ c. milk
¾ c. coconut, flaked
1 tsp. lemon peel, grated

Ingredients for glaze:

¼ c. powdered sugar
2 Tbs. lemon juice

Directions:

1. Preheat oven to 350 degrees F.
2. Lightly grease and flour an 8 x 4 x 2-inch loaf pan.
3. In large mixing bowl, cream butter, sugar, and lemon juice.
4. Add eggs, one at a time, beating well after each addition.
5. Combine flour, baking powder, and salt.
6. Add to creamed mixture alternately with milk.
7. Stir in coconut and lemon peel.
8. Bake 60 to 70 minutes, until inserted toothpick comes out clean.
9. Cool 10 minutes; turn out on wire rack.
10. In small bowl, combine glaze ingredients.
11. Brush over warm loaf.

Coconut Curry Bread

This is a recipe for bread to make with your bread machine. Coconut and peanuts add crunch to this bread seasoned with curry. Serve with your favorite fresh fruit salad.

Ingredients:

1 c. water
1 tsp. salt
1 Tbs. butter
3 c. bread flour
½ c. coconut, flaked
½ c. peanuts, roasted, coarsely chopped
3 Tbs. dry milk powder
1 Tbs. sugar
1½ tsp. curry powder
2 tsp. yeast

Directions:

1. Add ingredients to bread machine as suggested by manufacturer, adding coconut and peanuts with flour.
2. Bake on basic/white bread cycle; medium/normal.

Yields: 1 1½-lb. loaf

Coconut English Muffins

These are a delicious version of English muffins. They are great toasted and served with warm butter.

Ingredients:

1 c. warm water
1 pkg. dry yeast
1 tsp. sugar
2 tsp. salt

¼ c. butter
3 c. flour
1 c. coconut, flaked

Directions:

1. In large bowl, sprinkle yeast over water.
2. Mix in remaining ingredients.
3. On floured surface, roll out dough to ⅓-inch thickness; cut into circles.
4. Sprinkle wax paper with flour.
5. Allow to rise until doubled.
6. Bake on preheated griddle, over medium heat, until browned on both sides.

Coconut Buns

These make great coconut buns.

Ingredients:

1 c. butter
1 c. sugar
3 eggs
1½ c. flour
2 tsp. baking powder
½ lg. peeled coconut, grated
½ tsp. almond extract

Directions:

1. Preheat oven to 425 degrees F.
2. Lightly grease large muffin tins.
3. In large bowl, cream butter and sugar together.
4. Beat in eggs one at a time; stir in extract.
5. In small bowl, combine flour and baking powder.
6. Cut flour mixture into cream mixture.
7. Add grated coconut, mix to a soft dropping consistency.
8. Bake 20 to 30 minutes.

Orange Coconut Rolls

Try this recipe. These rolls are delicious.

Ingredients for dough:

1 pkg. yeast
¼ c. water, lukewarm
¼ c. sugar
1 tsp. salt
½ c. sour cream
6 Tbs. butter, melted
2 eggs
3 c. flour

Ingredients for filling:

¾ c. sugar
1 c. coconut, toasted
5 Tbs. grated orange peel, divided
2 Tbs. butter

Ingredients for glaze:

¾ c. sugar
½ c. sour cream
¼ c. butter
2 Tbs. orange juice

Directions for bread machine:

1. Place dough ingredients in bread machine in order recommended by the manufacturer.
2. Select dough cycle, press start.
3. When cycle completes, turn dough onto floured surface.
4. In small bowl, combine sugar, coconut, and orange peel.

5. Divide dough in half.
6. Roll out to a 12-inch circle.
7. Brush with 1 tablespoon butter.
8. Sprinkle with half of coconut mixture.
9. Cut into 12 wedges.
10. Roll up wedges starting with wide end.
11. Repeat with remaining dough.
12. Generously grease, not oil a 9 x 13-inch baking pan.
13. Place rolls point-side down in 3 rows in pan.
14. Let rise until doubled, 45 to 60 minutes.
15. Preheat oven to 350 degrees F.
16. Bake 25 to 30 minutes, until brown.
17. Leave in pan.
18. In small saucepan, combine glaze ingredients.
19. Boil 3 minutes, stirring occasionally.
20. Pour glaze over hot rolls.
21. Sprinkle with remaining toasted coconut.

Directions for manual method:

1. In mixer bowl, soften yeast in warm water.
2. Stir in sugar, salt, sour cream, butter, and eggs.
3. Add 1¼ cups flour.
4. Beat 2 minutes at medium speed.
5. Stir in remaining flour to form a soft dough.
6. Cool.
7. Let rise in warm place until doubled, about 45 minutes.
8. Knead dough about 15 times.
9. Proceed with above instructions at number 4.

Did You Know? . . . .

Did you know that a coconut tree is known as the tree of life? They are known as the tree of life because of their large variety of uses.

Pumpkin Coconut Bread

This is a wonderful bread to serve during the fall holidays. The coconut makes it moist and flavorful.

Ingredients:

2 c. sugar
1 c. brown sugar, packed
1 c. canola oil
4 eggs
1 can pumpkin (15 oz.)
3½ c. flour
2 tsp. baking soda
2 tsp. salt
½ tsp. ground cloves
1½ tsp. ground cinnamon
1 tsp. ground allspice
1 tsp. ground nutmeg
⅔ c. water
1½ c. coconut, flaked
1 c. walnuts, chopped

Directions:

1. Preheat oven to 350 degrees F.
2. Grease and flour two 9 x 5-inch loaf pans.
3. In large bowl, mix together sugars, oil, and eggs.
4. Mix in pumpkin.
5. Add flour, salt, soda, and spices, and then water.
6. Stir together until just moistened.
7. Stir in coconut and nuts.
8. Pour into prepared pans.
9. Bake 1 hour, until inserted toothpick comes out clean.
10. Cool 10 minutes.
11. Turn out onto wire rack.

Coconut Delights Cookbook

A Collection of Coconut Recipes

Cookbook Delights Series – Book 4

Breakfasts

Table of Contents

Page

Caribbean Brunch Strata

This is a tasty strata that is different than the usual.

Ingredients:

- 6 English muffins, halved, split
- ¼ tsp. dried hot peppers
- 1 boneless, skinless chicken breast, cut into ½-inch pieces
- 1 Tbs. honey
- 1 Tbs. olive oil
- 4 Tbs. coconut, flaked or shredded
- 5 oz. mild goat cheese, crumbled
- 6 eggs
- 3¾ c. milk
- ¼ c. green onions with tops, sliced
- ½ tsp. liquid red pepper seasoning
- ½ tsp. salt

Directions:

1. Butter a 2-quart glass casserole bowl.
2. Arrange half of English muffin pieces on bottom.
3. Sprinkle dried hot peppers and honey on chicken; toss to coat evenly.
4. In large skillet, heat oil over medium heat.
5. Add chicken; cook, stir until firm, 3 to 5 minutes.
6. Spoon over muffins.
7. Sprinkle with green onions, coconut, and goat cheese.
8. Top with remaining muffin pieces.
9. Beat eggs, milk, red pepper seasoning, and salt together until blended.
10. Ladle mixture over muffins.
11. Cover with plastic wrap and refrigerate several hours or overnight.
12. Preheat oven to 325 degrees F.

13. When ready to bake, remove plastic cover.
14. Bake 50 to 60 minutes, until puffy and brown.
15. If necessary, cover with foil the last 10 minutes of baking to prevent over-browning.
16. Note: Half a pound of peeled, deveined medium shrimp may be substituted for chicken.

Yields: 4 servings.

Carob-Coconut Pancakes

These pancakes are for those who are trying to add healthier ingredients to their pancakes. Try these coconut pancakes with the flavor of carob.

Ingredients:

2½ c. whole wheat pastry flour
2½ c. oat flour
1 Tbs. baking powder
½ c. raw carob powder
⅛ tsp. stevia powder
½ c. coconut, unsweetened, shredded
4 c. filtered water
olive oil spray

Directions:

1. In medium bowl, combine all dry ingredients.
2. Add 4 cups of water.
3. Mix until ingredients are combined well. You may need to add more water as the grains slowly absorb water while they sit in the bowl.
4. Heat skillet over medium heat.
5. Lightly grease skillet with spray.
6. Pour ¼ cupfuls of batter onto skillet.
7. Turn when tops form bubbles.
8. Serve with butter, maple syrup, or jam.

Coconut and Banana Pancakes

Here are some very rich pancakes for the hearty appetites. The strawberries and bananas are delicious.

Ingredients:

2 c. coconut milk
¼ c. dark brown sugar, firmly packed
1 egg
2 c. flour
1 lg. banana, finely chopped
1¾ oz. butter, melted
8 oz. fresh strawberries, chopped
⅓ c. maple syrup
sliced bananas
vanilla ice cream, to serve

Directions:

1. In large bowl, whisk milk, sugar, and egg until combined.
2. Gradually whisk in flour until smooth.
3. Stir in banana.
4. Heat a nonstick frying pan over a medium heat.
5. Brush pan with melted butter.
6. Pour ¼ cup of batter into pan.
7. Cook, uncovered, until bubbles appear on surface of pancake.
8. Turn pancake, cook until browned.
9. Cover to keep warm.
10. Repeat with remaining butter and batter.
11. Blend strawberries with maple syrup until smooth.
12. Serve pancakes layered with sliced bananas.
13. Top with vanilla ice cream and drizzle with strawberry sauce.

Yields: 4 pancakes.

Coconut Crumpets

Our family loves crumpets and we serve them warm with jam or jelly. Try this and enjoy with your favorite tea.

Ingredients:

1 pkg. instant yeast or 1½ tsp. yeast
3 c. flour
1 c. coconut, shredded
1 tsp. sugar
1 c. milk
1¾ c. water
½ tsp. salt
½ tsp. bicarbonate of soda

Directions:

1. Preheat oven to 400 degrees F.
2. In medium bowl, sift yeast, flour, and sugar.
3. In small saucepan, heat milk and water.
4. Whisk milk mixture gradually into flour mixture until a smooth batter is formed; cover.
5. Turn off oven; put mixture in oven and let stand for one hour or until doubled in volume.
6. Stir in salt and bicarbonate of soda.
7. Heat heavy pan until a small piece of butter browns and burns within 10 seconds; wipe burnt butter off pan.
8. Make sure batter is consistency of pouring cream.
9. If too thin, thicken with a little extra flour. If it is too thick, thin with a little warm water.
10. Place egg rings, biscuit cutters, or muffin rings onto pan.
11. Spray lightly with cooking spray to coat.
12. Spoon 2 to 3 tablespoons mixture into rings or cutters and cook until golden brown on each side.
13. Serve warm with your favorite jam and tea.

Coconut Cherry Scones with Citrus Butter

Serve these tender, warm scones along with delightful citrus butter for a mid-morning break.

Ingredients for butter:

½ c. butter, softened
1 Tbs. powdered sugar
1 tsp. freshly grated lemon peel
1 tsp. freshly grated orange peel

Ingredients for scones:

2 c. flour
¼ c. sugar
2½ tsp. baking powder
¼ tsp. salt
½ c. cold butter
1 egg, beaten
½ c. half and half cream
⅓ c. coconut, sweetened, flaked
½ c. dried cherries or cranberries, chopped
1 tsp. lemon peel, freshly grated
1 Tbs. powdered sugar

Directions for butter:

1. In small bowl, combine all ingredients.
2. Beat at low speed, until creamy; set aside.

Directions for scones:

1. Preheat oven to 375 degrees F.
2. In medium bowl, combine flour, sugar, and baking powder.
3. Cut in ½ cup butter with pastry blender or fork until mixture resembles coarse crumbs.
4. In a small bowl, combine egg, half and half, coconut, cherries, and lemon peel.

5. Add to flour mixture.
6. Stir just until flour mixture is moistened.
7. Turn dough onto lightly floured surface; knead lightly 8 to 10 times.
8. Pat dough into 7-inch circle.
9. Place onto greased baking sheet.
10. Cut into 8 wedges, do not separate.
11. Sprinkle with powdered sugar.
12. Bake 20 to 25 minutes, until golden brown.
13. Cool 15 minutes.
14. Cut wedges apart; remove from baking sheet.
15. Serve warm scones with citrus butter.

Coconut Milk Pancakes

Try these coconut-flavored pancakes for a change of pace.

Ingredients:

2 Tbs. coconut oil or butter, melted
¼ tsp. coconut or vanilla extract
½ c. coconut milk
1 tsp. sugar
⅛ tsp. salt
3 Tbs. coconut flour, sifted
⅛ tsp. baking powder
2 eggs

Directions:

1. In a bowl, blend together eggs, oil, extract, coconut milk, sugar, and salt.
2. In another bowl, combine flour and baking powder; mix thoroughly.
3. Combine dry ingredients with wet to form a smooth batter.
4. Heat 1 tablespoon of coconut oil on griddle or skillet.
5. Spoon batter onto hot surface, making pancakes about 2½ to 3-inches in diameter.

Coconut Orange Scones

Orange and coconut add great flavor to these scones.

Ingredients:

3¼ c. plus 2 Tbs. flour
1¾ Tbs. sugar, plus ¼ c. for topping
½ c. fresh coconut, grated
2 tsp. baking powder
4 Tbs. butter, softened
2 eggs
1 c. cold milk, plus 1 Tbs. for glaze
1 lg. orange, zested
1 egg yolk
pinch sea salt
honey and butter, for serving

Directions:

1. Preheat oven to 300 degrees F.
2. Line large baking tray with parchment paper; set aside.
3. In large bowl, sift flour, salt, sugar, and baking powder.
4. Cut butter into small cubes and rub into flour mixture with your hands.
5. In medium bowl, lightly beat the eggs.
6. Add milk and mix.
7. Pour egg mixture into flour mixture.
8. Mix until just blended.
9. Add 3 teaspoons orange zest and lightly knead dough. It is important to keep the kneading to a minimum, or the scones will be hard and tough.
10. Put dough onto lightly floured surface.
11. Roll to 1-inch thickness.
12. Cut dough with 2-inch round fluted cutter.
13. Put dough circles on baking sheet, 1 inch apart.

14. In small bowl, lightly beat egg yolk and remaining milk.
15. Brush over scone tops.
16. In another bowl, mix remaining sugar, coconut, and orange zest.
17. Sprinkle evenly over top of each scone.
18. Bake in center of oven, 20 minutes, until light brown.
19. Cool on wire rack.
20. Split open, serve with butter and/or honey.

Coconut Orange Pancakes

This recipe is a lovely change from traditional pancakes.

Ingredients:

- 1 c. whole wheat flour
- 1½ tsp. baking powder
- ¼ tsp. salt
- ¼ c. coconut, grated
- 1 Tbs. molasses
- 1 egg
- ¼ c. coconut oil
- 1¼ c. orange juice, lukewarm

Directions:

1. In small bowl, mix coconut, baking powder, salt, and flour.
2. In another bowl, mix egg, molasses, coconut oil, and orange juice in a separate container.
3. Lightly grease skillet with coconut oil.
4. Add dry ingredients to liquid ingredients.
5. Mix gently.
6. Spoon by tablespoons into hot skillet to make 3-inch pancakes.

Coconut Scones

Here is a version of scones that you will enjoy.

Ingredients:

- 2 c. flour
- 1 Tbs. baking powder
- 1 tsp. nutmeg
- 1¾ c. butter
- 3 tsp. sugar
- 1 c. coconut, dessicated
- 1 c. milk
- 1 egg, lightly beaten

Directions:

1. Preheat oven to 400 degrees F.
2. Lightly grease baking sheet.
3. In large bowl, sift dry ingredients together.
4. Add ½ cup coconut.
5. Blend in butter until mixture resembles fine bread crumbs.
6. In small bowl, add egg and milk; mix together.
7. Make a well in center of flour mixture.
8. Reserving a small amount of egg mixture, pour into flour mixture.
9. Mix lightly and quickly.
10. Turn out onto floured surface, knead lightly.
11. Pat or roll dough out to ¾-inch thickness.
12. Cut into rounds with 2-inch cutter dipped into flour each time before cutting.
13. Brush each round with lightly beaten egg and milk.
14. Dip in extra coconut.
15. Place close together on lightly greased scone tray.
16. Let rest 10 minutes.
17. Bake 10 minutes, until golden brown.

Crunchy Coconut French Toast

This is another version of French toast for a change of pace.

Ingredients:

1¼ c. coconut, unsweetened, shredded
¼ c. milk
1 tsp. vanilla extract
3 c. cornflakes, slightly crushed
4 lg. eggs
4 croissants, sliced lengthwise or 4 pieces thick bread
1 Tbs. butter, unsalted
maple syrup, for serving

Directions:

1. In shallow bowl, add coconut and cornflakes; mix well.
2. In another shallow bowl, whisk together eggs, milk, and vanilla.
3. Dip bread slices into egg mixture and soak for a minute on each side. They should be well coated but not soggy.
4. Press each slice into coconut mixture on both sides, patting firmly; turning them over several times to coat thoroughly.
5. Heat butter, on griddle or large pan, over medium heat.
6. Cook 2 to 3 minutes on each side, until golden brown.
7. Serve with warm maple syrup.
8. Note: It is important to use unsweetened coconut in this recipe.
9. The sweetened variety is too moist and makes the bread soggy.

Coconut Belgian Waffles

Here is a version of waffles that is low fat.

Ingredients:

1½ c. flour
2 tsp. baking powder
½ c. coconut, shredded
1 c. skim milk
½ c. fat-free-egg-substitute
2 Tbs. maple syrup
2 Tbs. honey
6 egg whites
4 Tbs. canola oil

Directions:

1. Preheat waffle iron.
2. In large bowl, whisk flour, baking powder, and coconut.
3. In small bowl, whisk syrup, honey, and egg substitute.
4. Stir in oil and milk.
5. In medium bowl, beat egg whites until stiff.
6. Make a hole in center of dry ingredients.
7. Pour liquid mixture into hole.
8. Gently fold whites into batter.
9. Bake until golden brown, 6 to 8 minutes.

Coconut Popovers

Our family loves popovers, and this coconut version is also very good.

Ingredients:

1 c. milk
3 eggs
1 Tbs. olive oil
½ tsp. salt

1 c. flour, sifted
¾ c. coconut, shredded

Directions:

1. In blender, place milk, eggs, oil, and salt.
2. Blend at high speed until well mixed.
3. Add flour to mixture in blender container.
4. Add coconut.
5. Blend just until batter is smooth.
6. Pour batter into greased popover cups, filling half full.

Coconut Pecan French Toast

This is a fancy-tasting breakfast treat that is really very easy to make.

Ingredients:

1¼ c. milk
1½ c. coconut, flaked
½ c. pecan halves, chopped
1½ tsp. vanilla extract
2 Tbs. praline extract
2 Tbs. maple syrup
6 lg. eggs
12 slices white bread

Directions:

1. In large bowl, mix eggs, milk, extracts, and syrup together.
2. In large shallow bowl, mix coconut and pecans together; set aside.
3. Dip each bread slice into egg mixture.
4. Then dip only one side of bread into the coconut mixture.
5. Spray or oil skillet or griddle on medium heat.
6. Cook both sides until the egg is set.
7. Serve with maple syrup, if desired.

Coconut Waffles with Berry Mango Compote

Flavored with coconut and mango, these waffles take on a tropical note. These are also good with raspberries or any berries you have on hand.

Ingredients for compote:

1 ripe mango, peeled, pitted, cut into small cubes
1 pt. strawberries, hulled, sliced
1 Tbs. sugar

Ingredients for waffles:

1¾ c. flour
2 Tbs. sugar
1 Tbs. baking powder
½ tsp. salt
½ c. coconut milk
¼ c. coconut, sweetened, shredded
8 Tbs. butter, unsalted, melted
1 c. milk
3 lg. eggs, beaten
fresh mint, for garnish

Directions for compote:

1. In medium bowl, toss mango, strawberries, and sugar; cover.
2. Let stand until fruit gives off some juices, about 90 minutes.

Directions for waffles:

1. Lightly oil waffle iron grids; preheat.
2. In large bowl, combine flour, sugar, baking powder, and salt.
3. Make a well in center; pour in melted butter, coconut milk, eggs, and milk.

4. Whisk until smooth; do not over mix.
5. Mix in coconut.
6. Spoon ¼ cup of batter into center of each quadrant of waffle iron.
7. Cook until golden brown, 3 to 4 minutes.
8. Serve with compote on top.
9. Garnish with mint leaves.
10. Note: Let fruit compote stand at room temperature for at least thirty minutes before serving.

Coconut Macadamia Nut French Toast

The chopped macadamia nuts and coconut add great flavor to the usual French toast.

Ingredients:

4 1½-inch slices sweet bread, cut in half
½ c. coconut, shredded
½ c. cornflake crumbs
¼ c. macadamia nuts, finely chopped
½ c. milk
4 eggs, lightly beaten

Directions:

1. In shallow bowl, combine coconut, crumbs, and macadamia nuts.
2. In medium bowl, whisk eggs and milk until well blended.
3. Dip, do not soak, bread slices in egg mixture, coating both sides evenly.
4. Press one side of dipped bread into coconut mixture.
5. Spray or oil skillet and preheat.
6. Cook coconut side of bread first until golden brown, turn; cook other side until golden brown and egg is set.
7. Repeat with remaining bread slices; serve with warm syrup or whipped cream, if desired.

Coconut Oatmeal

Coconut and walnuts add great flavor and more texture to old fashioned oatmeal. Enjoy!

Ingredients:

- 3½ c. milk
- 2 c. old-fashioned oats
- 1 c. apple, unpeeled, chopped
- ¼ c. maple syrup
- ¾ c. coconut
- ¾ c. walnuts, chopped
- 1 Tbs. butter

Directions:

1. In medium saucepan, bring milk to gentle boil.
2. Stir in oats.
3. Add butter and salt.
4. As soon as it begins to boil again, reduce heat to medium.
5. Cook for amount of time directed on oatmeal package.
6. Stir often.
7. Remove from heat.
8. Add maple syrup, coconut, walnuts, and apple.
9. Mix well.
10. Cover; let sit 3 minutes.
11. Serve with additional maple syrup, a little milk and sprinkle with brown sugar and more coconut.

Yields: 4 servings.

Did You Know? . . . .

Did you know that a rib of a branch of the coconut tree can be used to make fire tongs?

Cakes

Table of Contents

Page

Apple Coconut Cake

The addition of coconut and many nuts make this a moist and flavorful cake full of texture and flavor.

Ingredients:

- 3 eggs
- 2 c. sugar
- 1½ c. canola oil
- 3 c. flour
- 1 tsp. baking powder
- 1 tsp. baking soda
- 2 tsp. vanilla extract
- 1½ c. apples, peeled, cored, cut into pieces
- 1¾ c. coconut
- 1¾ c. walnuts, chopped

Directions:

1. Preheat oven to 350 degrees F.
2. Grease a 9 x 13-inch baking pan.
3. In large bowl, beat eggs; add sugar and vanilla.
4. In another bowl, sift flour, baking powder, baking soda, and salt.
5. To large bowl, add sifted ingredients, combine well.
6. Add oil, apples, coconut, and walnuts; combine well.
7. Pour into prepared pan.
8. Bake 40 minutes, until inserted toothpick comes out clean.
9. Sprinkle with powdered sugar after it is cooled.

Did You Know?. . . .

Did you know that the fibers from coconut husks make mats, mattresses, mosquito repellant, and rope?

Coconut Lemon Loaf

This is an easy-to-make cake-style pound cake. Make two while you are at it because it freezes well.

Ingredients for glaze:

¼ c. powdered sugar
1 Tbs. lemon juice

Ingredients for loaf:

½ c. butter, softened
4½ tsp. lemon juice
1½ c. flour
¼ tsp. salt
½ c. milk
½ c. coconut, flaked
1 c. sugar
1 tsp. baking powder
2 eggs
1 tsp. lemon peel, grated

Directions:

1. Preheat oven to 350 degrees F.
2. Lightly grease an 8 x 4 x 2-inch loaf pan.
3. In mixing bowl, cream butter, sugar, and lemon juice.
4. Add eggs, one at a time, beating well after each addition.
5. Combine flour, baking powder, and salt.
6. Add to creamed mixture alternately with milk.
7. Stir in coconut and lemon peel.
8. Pour into prepared pan.
9. Bake 60 to 70 minutes, until inserted toothpick comes out clean.
10. Cool 10 minutes.
11. Remove from pan to wire rack.
12. Combine glaze ingredients and brush over loaf.

Chocolate Cake with Toasted Almond Coconut Topping

This cake is moist and flavorful.

Ingredients for cake:

1½ c. coconut, flaked or shredded
1½ c. sliced almonds
3 c. sugar
2¾ c. flour
1¼ c. cocoa
2¼ tsp. baking powder
2¼ tsp. baking soda
1½ tsp. salt
1½ c. milk
¾ c. canola oil
1 Tbs. vanilla extract
1½ c. very hot water
3 eggs

Ingredients for topping:

1½ c. sugar
⅓ c. water
1½ c. heavy cream

Directions for cake:

1. Preheat oven to 350 degrees F.
2. Spread coconut and almonds on separate baking sheets and toast, 10 to 15 minutes, stirring occasionally, until golden brown.
3. Cool.
4. Line bottom and sides of a 14 x 17-inch baking sheet with parchment.
5. Sift together dry ingredients.
6. Put in mixer bowl and blend briefly.

7. In medium bowl, whisk eggs, milk, oil, and vanilla.
8. Add to dry ingredients.
9. Mix at low speed for 5 minutes.
10. Gradually add hot water; mix until just combined.
11. Pour batter into prepared pan.
12. Bake 25 to 30 minutes, until inserted toothpick comes out clean.
13. Cool in pan.
14. Cut cake into triangles and pour sauce over.

Directions for sauce:

1. Pour sugar into deep saucepan.
2. Carefully pour water around sugar; do not splash sugar onto sides of pan.
3. Do not stir.
4. Gently draw your finger through center of sugar to moisten.
5. Over medium-high heat, bring to boil without stirring.
6. Reduce heat to a fast simmer.
7. Cook 10 to 20 minutes, without stirring, until amber-caramel in color.
8. Immediately remove from heat.
9. Using a wooden spoon, slowly and carefully stir in cream, it will bubble up and splatter.
10. Stir in toasted coconut and almonds; set aside.
11. Note: If sauce has stiffened, rewarm it over low heat or in microwave until softened.

Did You Know? . . . .

Did you know coconuts apparently kill approximately 150 people each year by falling from 80 feet high? A coconut can build up an impact speed of 50 miles per hour.

Coconut Apple Pound Cake

This makes a very moist and delicious cake. This also freezes well, so you may want to make two while you are at it.

Ingredients for cake:

1 tsp. vanilla extract
1½ c. sugar
1½ c. canola oil
3 eggs
3 c. flour
1 tsp. baking soda
1 tsp. salt
3½ c. apple, chopped
1½ c. coconut
1½ c. nuts, chopped

Ingredients for glaze:

¾ c. sugar
½ tsp. baking soda
½ c. buttermilk
½ c. butter
1 tsp. vanilla extract, boil 5 minutes
1 Tbs. corn syrup

Directions for cake:

1. Preheat oven to 375 degrees F.
2. Lightly grease a bundt pan.
3. In large bowl, blend first 7 ingredients with mixer.
4. Stir in apples, coconuts, and nuts.
5. Pour into prepared pan.
6. Bake 1 hour, 20 minutes.
7. Cool.
8. Turn out of pan onto serving plate.
9. With a fork, punch small holes in top of cake.
10. Pour glaze over cake.

Directions for glaze:

1. In small saucepan, combine all ingredients.
2. Bring to boil; boil 5 minutes.

Blueberry Coconut Pound Cake

This recipe can be made as a pound cake or muffins. It is moist and delicious.

Ingredients:

½ c. butter, softened
¾ c. sugar
2 tsp. lime zest, freshly grated
5 Tbs. heavy cream
¾ c. blueberries
¾ c. coconut, sweetened, flaked
1 c. flour
2 lg. eggs

Directions:

1. Preheat oven to 350 degrees F.
2. Put oven rack in middle.
3. Grease or paper-line 7 to 9 half-cup muffin cups.
4. In a mixer bowl, beat butter, sugar, and zest until light and fluffy.
5. Beat in eggs, one at a time.
6. Add cream and flour.
7. Mix on low speed until just combined.
8. Stir in ½ cup of coconut.
9. Gently stir in blueberries.
10. Spoon batter into cups, filling the cups.
11. Smooth tops.
12. Sprinkle tops with remaining coconut.
13. Bake 25 minutes, until inserted toothpick comes out clean and edges are golden brown.
14. Invert on wire rack and cool.

Coconut Banana Cake

If you love bananas and coconut, you will love this cake. Enjoy!

Ingredients for cake:

1½ c. sugar
½ c. buttermilk
1½ c. banana, ripe, mashed
1½ c. cake flour
½ tsp. salt
¾ tsp. baking soda
¾ c. pecans, finely chopped
1 tsp. baking powder
3 lg. eggs
8 oz. butter

Ingredients for filling:

⅔ c. sugar
⅓ c. cornstarch
1 can unsweetened coconut milk
1 c. milk
5 egg yolks
¼ c. lemon juice
3 bananas, sliced
2 Tbs. sugar
1½ c. coconut, unsweetened, flakes, toasted

Directions for cake:

1. Preheat oven to 350 degrees F.
2. Grease and flour two 9-inch baking pans.
3. In mixer bowl, beat butter and sugar until light and fluffy.
4. Add eggs, one at a time, beating well after each addition.
5. Add buttermilk and mashed banana; mix well.
6. In separate bowl, sift dry ingredients.

7. Add to butter mixture; mix at low speed until moistened.
8. Beat one minute at medium speed.
9. Fold in pecans.
10. Pour batter evenly into prepared pans.
11. Bake 25 to 35 minutes, until inserted toothpick comes out clean.
12. Remove from oven; cool 5 minutes.
13. Invert onto a wire rack; cool completely.

Directions for filling:

1. Line a baking sheet with parchment paper.
2. Cut another piece the same size.
3. In a bowl combine coconut milk and milk.
4. Stir together.
5. In a bowl, mix sugar and cornstarch.
6. Add egg yolks and coconut milk to sugar mixture.
7. Whisk until smooth.
8. In heavy saucepan, bring the remaining coconut milk mixture to a boil.
9. Pour into egg yolk-sugar mixture, whisking constantly.
10. Return mixture to saucepan over medium heat and continue stirring.
11. Cook until small bubbles start to form on the top.
12. Remove from heat.
13. Spread onto lined baking sheet.
14. Cover top of the filling with the other sheet of parchment.
15. Cool in refrigerator.
16. In large bowl, gently toss banana slices with lemon juice and sugar.
17. Trim top of first cake so that surface is level.
18. Spread thin layer of custard.
19. Place bananas completely over bottom layer.
20. Spread another layer of filling.
21. Place next cake layer on top and spread the top generously with filling.
22. Garnish with toasted coconut flakes.

Coconut Cherry Cheese Coffee Cake

Cherries and coconut add flavor to this easy-to-make coffee cake.

Ingredients:

- 2½ c. flour
- ¾ c. sugar
- ½ tsp. baking powder
- ½ tsp. baking soda
- 2 pkg. cream cheese, softened, divided (3 oz. each)
- ¾ c. milk
- 2 Tbs. olive oil
- 1 tsp. vanilla extract
- 1 c. coconut, flaked
- 2 eggs
- ¾ c. cherry preserves
- 2 Tbs. butter

Directions:

1. Preheat oven to 350 degrees F.
2. Grease and flour a 9-inch springform pan.
3. In large bowl, combine flour and sugar.
4. Reserve ½ cup flour mixture; set aside.
5. Stir baking powder and soda into remaining flour mixture.
6. Cut in 1 package cream cheese with pastry blender until mixture resembles coarse crumbs; set aside.
7. In medium bowl, combine milk, oil, and 1 egg.
8. Add to cream cheese mixture.
9. Stir until just moistened.
10. Spread batter in prepared pan.
11. Sprinkle coconut over cream cheese mixture.
12. Spoon preserves evenly over coconut.
13. Cut butter into reserved flour mixture with pastry blender.

14. Sprinkle preserves over top.
15. Bake 55 to 60 minutes, until inserted toothpick comes out clean.
16. Cool in pan on wire rack, 15 minutes.
17. Remove from pan.
18. Serve warm.

Coconut Chocolate Frosting

This makes a delicious chocolate frosting.

Ingredients:

- ⅓ c. sugar
- ¾ c. evaporated milk
- ¾ c. nuts, chopped
- 1 Tbs. cornstarch
- 2 Tbs. butter
- 1½ c. coconut flakes, sweetened
- 3 milk chocolate bars, broken into pieces (1.55 oz.)

Directions:

1. In small saucepan, stir together sugar and cornstarch.
2. Stir in evaporated milk.
3. Cook over medium heat, stirring constantly, until mixture boils.
4. Remove from heat.
5. Add chocolate and butter.
6. Stir until chocolate is melted and mixture is smooth.
7. Stir in coconut and nuts.
8. Immediately spread on cake.

Yields: 2 cups.

Coconut Pineapple Upside-Down Cake

This is always a fun cake to make, serve, and eat.

Ingredients:

⅔ c. butter
¾ c. brown sugar, lightly packed
1 sm. can pineapple, crushed
1½ c. pecans, chopped
1½ c. coconut, flaked
1 sm. jar maraschino cherries, drained
1⅓ c. flour
1 tsp. baking powder
½ tsp. salt
¼ tsp. baking soda
⅔ c. buttermilk
1 egg
1 tsp. vanilla extract
1 c. sugar

Directions:

1. Preheat oven to 350 degrees F.
2. Melt ⅓ cup butter in a 9-inch square pan.
3. Sprinkle brown sugar over butter.
4. Drain pineapple, spread over brown sugar.
5. Spread pecans, coconut, and cherries over pineapple.
6. In medium bowl, sift dry ingredients together.
7. Cut in remaining butter.
8. Add buttermilk, egg, and vanilla; beat until smooth.
9. Pour over pineapple mixture.
10. Bake 45 minutes.
11. Turn upside down on plate.
12. Let stand a few minutes.
13. Remove pan and cool.

Coconut Pound Cake

My husband loves pound cake, and this is a very flavorful and moist cake.

Ingredients for cake:

1½ tsp. baking powder
½ c. milk
3½ oz. coconut, flaked
2 c. sugar
1 c. butter
6 eggs
2 c. flour
1 tsp. coconut extract

Ingredients for glaze:

¼ c. water
1 c. sugar
1 tsp. coconut extract
1 c. walnuts, chopped

Directions:

1. Preheat oven to 350 degrees F.
2. Grease and flour a 10-inch tube pan.
3. In large bowl, cream sugar and butter.
4. Fold in eggs one at a time, mixing well after each addition.
5. In separate bowl, mix flour, baking powder, milk, coconut, and extract.
6. Beat into the creamed mixture.
7. Pour batter into prepared pan.
8. Bake 1 hour, until inserted toothpick comes out clean.
9. Cool on a wire rack for a few minutes.
10. Place on cake platter.
11. In small saucepan, combine glaze ingredients.
12. Boil 1 minute; pour over warm cake.

German Sweet Chocolate Cake

This cake has always been one of our family favorites. The cake is moist and the frosting is delicious. Enjoy!

Ingredients for cake:

2½ c. flour
½ c. boiling water
¼ tsp. salt
1 pkg. German sweet chocolate (4 oz.)
1 c. butter
2 c. sugar
4 egg yolks
1 tsp. vanilla extract
1 tsp. baking soda
1 c. buttermilk
4 egg whites, stiffly beaten

Ingredients for frosting:

1 c. evaporated milk
1 c. sugar
1 Tbs. butter
1½ c. coconut, flaked
1 c. pecans, chopped
1 tsp. vanilla extract
3 egg yokes

Directions for cake:

1. Preheat oven to 350 degrees F.
2. Grease and flour three baking pans.
3. Melt chocolate in boiling water; cool.
4. In large bowl, cream butter and sugar until fluffy.
5. Add yolks, beating well after each.
6. Blend in vanilla and chocolate.
7. In small bowl, sift flour with soda and salt.
8. Add dry mixture alternately with buttermilk, mixing well after each addition.

9. Fold in beaten egg whites.
10. Pour into prepared pans.
11. Bake 35 to 40 minutes.

Direction for frosting:

1. In medium saucepan, beat eggs, milk, sugar, and butter.
2. Over medium heat, cook 12 minutes, stirring until thickened; remove from heat.
3. Add coconut, pecans, and vanilla.
4. Beat until cool and of spreading consistency.
5. Frost between layers, top, and sides.

Coconut Pecan Frosting

This is a delicious and rich frosting that is great to add to your favorite cake or cupcakes.

Ingredients:

1½ c. evaporated milk
1½ c. sugar
¾ c. butter
1½ tsp. vanilla extract
2 c. coconut
1¾ c. pecans, chopped
4 egg yolks, slightly beaten

Directions:

1. In saucepan, combine all ingredients except coconut and pecans.
2. Cook and stir over medium heat until thickened.
3. Remove from heat.
4. Stir in coconut and pecans.
5. Cool until thick enough to spread.

Yields: 4½ cups.

Coconut Crumb Coffee Cake

The crumb topping is a nice addition to this coffee cake.

Ingredients:

- 4 c. flour
- 1 tsp. baking powder
- ⅔ c. butter
- ½ tsp. salt
- 2½ c. sugar
- 2 eggs, beaten
- 1 tsp. baking soda
- 1 tsp. cinnamon
- 1½ c. buttermilk
- 1 c. brown sugar
- 1 c. coconut
- ice cream or whipping cream (optional)

Directions:

1. Preheat oven to 350 degrees F.
2. Lightly grease and flour three 8-inch baking pans.
3. In medium bowl, blend first 5 ingredients together.
4. Set aside one cup of this mixture for topping.
5. In separate bowl, mix eggs, buttermilk, and baking soda together.
6. Blend remaining portion of crumb mixture, stirring thoroughly.
7. Add brown sugar, coconut, and cinnamon to the cup of reserved topping.
8. Pour coffee cake batter into prepared pans.
9. Sprinkle topping equally on top of batter in all three pans.
10. Bake 35 minutes.
11. Cool in pan 10 minutes.
12. Set on wire rack to cool completely.
13. Serve with ice cream or whipping cream.

Coconut Delights Cookbook

A Collection of Coconut Recipes
Cookbook Delights Series – Book 4

Candies

Table of Contents

Page

Chinese Coconut Candy

This recipe makes a simple candy that is very delicious.

Ingredients:

1 c. fresh coconut, grated
2 c. brown sugar
walnut halves

Directions:

1. Butter an 8-inch square pan.
2. In heavy saucepan, melt sugar over low heat.
3. Add coconut and cook until sugar begins to crystallize.
4. Remove from heat; pour into pan.
5. Cut into 1-inch squares.
6. Press a walnut half on top of each square.

Coconut Sprinkles

Use to decorate frosted cakes or cookies. Store sprinkles in an airtight container for future use.

Ingredients:

2 c. unsweetened coconut, flaked
½ c. maple syrup

Directions:

1. Preheat oven to 200 degrees F.
2. In medium bowl, combine both ingredients.
3. Bake until dry, stirring occasionally and taking care not to burn.

Chocolate Coconut Cherry Balls

These candy balls are actually very easy to make and are full of flavor and texture. They are delicious as a special treat or as a special gift.

Ingredients:

- ⅔ c. peanut butter, room temperature
- ¾ c. walnuts, chopped
- ½ tsp. salt
- ¼ bar paraffin wax (1 oz.)
- 4 Tbs. butter, room temperature
- 2 c. powdered sugar
- 1¼ c. coconut, shredded
- 24 maraschino cherries, chopped
- 1 pkg. semi-sweet chocolate chips (12 oz.)

Directions:

1. In large bowl, combine butter, powdered sugar, maraschino cherries, peanut butter, walnuts, coconut, and salt.
2. Shape into 1-inch balls.
3. In top of a double boiler, over hot water, melt chocolate chips and paraffin wax.
4. Insert a toothpick into each ball.
5. Dip into chocolate mixture to coat exterior.
6. Place onto wax paper to set.
7. Use another toothpick to push ball from inserted toothpick.
8. Note: If a hole remains where toothpick was inserted, add a small amount of chocolate coating to cover.

Yields: 3 dozen.

Chocolate Coconut Cherry Fudge

This is a delicious-tasting variation of fudge. It is delicious as a special treat on a holiday or as a gift for a special friend.

Ingredients:

1 can sweetened condensed milk (14 oz.)
2 Tbs. butter
⅔ c. vanilla chips (4 oz.)
2 c. semi-sweet chocolate chips (12 oz.)
¾ c. dried pitted tart cherries, chopped
¾ c. coconut, sweetened

Directions:

1. Line a 9-inch square pan with foil, extending above pan on two sides.
2. In medium saucepan, stir milk and butter over low heat until butter melts.
3. Remove from heat.
4. Measure ⅓ cup of mixture into small bowl.
5. Add vanilla chips and stir until melted.
6. Stir in cherries.
7. Add chocolate chips to mixture in saucepan.
8. Stir until melted, returning to heat if necessary.
9. Using half vanilla mixture, drop small teaspoonfuls into prepared pan, leaving spaces between each.
10. Sprinkle with half the coconut.
11. Spread chocolate mixture evenly on top.
12. Spoon on remaining vanilla mixture.
13. Sprinkle with remaining coconut.
14. Refrigerate 8 hours until firm enough to cut.
15. Lift foil by ends onto cutting board.
16. Cut fudge into 1-inch squares.

Coconut Almond Candy

This candy is melt-in-your-mouth delicious and is easy to make.

Ingredients:

¾ c. liquid coconut oil
1 c. almonds, chopped
1 c. coconut, dehydrated

Directions:

1. In small bowl, stir together oil, almonds, and coconut reserving ⅓ cup.
2. Pour into an 8 x 8-inch pan, sprinkle the top with coconut.
3. Place in refrigerator to chill for 1 hour.
4. Cut into bite-size squares to serve.
5. Store in refrigerator.

Trail Mix

Experiment with your favorite nuts and dried fruits to come up with your own favorite trail mix.

Ingredients:

1 c. raisins
1 c. coconut
1 c. sunflower seeds, shelled
1 c. semi-sweet chocolate chips
1 c. dried apricots, chopped
1 c. nuts, select your favorite

Directions:

1. In medium bowl, combine all ingredients; stir well.
2. Store in container with lid.

Coconut and Chocolate Candy

This is very easy and quick to make. Place in decorated mini muffin liners.

Ingredients:

½ c. butter, melted
2 c. powdered sugar
3 c. coconut, sweetened, flaked
2 oz. semi-sweet or milk chocolate chips, melted
almond halves or slivered almonds

Directions:

1. In large bowl, combine butter, sugar, and coconut; mix well.
2. Line baking sheet with wax paper.
3. Shape rounded teaspoonfuls into balls.
4. Place on baking sheet.
5. Flatten each ball slightly with finger.
6. Fill centers with melted chocolate.
7. Place an almond half or sliver on top, if desired.
8. Chill until firm.
9. Store in refrigerator.

Yields: 36 candies.

Coconut Apricot Balls

These are simple and easy to make and tastes great!

Ingredients:

1½ c. dried apricots, finely chopped
2 c. coconut, shredded
⅔ c. sweetened condensed milk
powdered sugar

Directions:

1. In large bowl, mix coconut and apricots.
2. Add condensed milk; blend well.
3. Shape into balls.
4. Roll in powdered sugar.
5. Let stand until firm.

Yields: 32 balls.

Coconut Candy Balls

These are delicious coconut candies that are very easy to make.

Ingredients:

1 c. butter, melted
1 can condensed milk
3½ c. coconut
2 lb. powdered sugar
1 pkg. chocolate chips (12 oz.)
1 pc. paraffin wax (2-inch sq.)

Directions:

1. In medium bowl, combine first 4 ingredients in order given.
2. Shape into balls.
3. Place on wax paper on baking sheet, let harden.
4. Chill.
5. On top of double boiler, melt chocolate chips and paraffin.
6. Keeping chocolate mixture on stove over water, dip balls into chocolate mixture.
7. Place on wax paper to cool, or place in wax paper candy cups.

Coconut Brittle

This makes a great candy brittle. Enjoy!

Ingredients:

2 c. sugar
1 c. white corn syrup
¼ c. water
2 c. coconut
1 tsp. vanilla extract
3 tsp. baking soda

Directions:

1. In 2-quart saucepan, add sugar, syrup, and water.
2. Heat to 225 degrees F.
3. Add coconut, cook to 300 degrees F.
4. Remove from heat.
5. Add vanilla and baking soda.
6. Pour onto buttered sheet or shallow pan.
7. When cold break into bite-size pieces.
8. Store in covered container.

Coconut Candy

This makes an easy-to-make candy for the coconut enthusiast.

Ingredients:

½ c. white corn syrup
½ tsp. vanilla extract
2 c. coconut
1 c. miniature marshmallows
1 pkg. milk or dark chocolate chips, melted (12 oz.)

Directions:

1. In medium bowl, combine coconut and vanilla.
2. Melt marshmallows in corn syrup.
3. Pour over coconut mixture and stir.
4. Roll into desired size balls.
5. Serve some plain or dipped in your favorite melted chocolate.

Coconut Fudge

Try this delicious coconut fudge for a change of pace.

Ingredients:

6 oz. cream cheese, softened
4½ c. powdered sugar
1½ c. blanched almonds, chopped
1½ c. coconut, flaked
½ tsp. coconut extract
sliced almonds

Directions:

1. Butter an 8-inch square pan.
2. In medium bowl, with electric mixer, beat cream cheese and enough powdered sugar to make a stiff but not dry mixture.
3. Stir in chopped almonds, coconut, and extract.
4. Press mixture into prepared pan.
5. Arrange sliced almonds on top.
6. Gently press them into the fudge.
7. Score fudge into squares with knife.
8. Refrigerate until firm.
9. Cut into squares.
10. Store in refrigerator.

Coconut Blonde Fudge

This fudge is delicious and the coconut and pecans give it excellent flavor. It also makes a great gift for a special person or to make for a special holiday.

Ingredients:

3 c. sugar
¾ c. milk
3 Tbs. light corn syrup
⅛ tsp. salt
3 Tbs. butter
2 tsp. vanilla extract
½ c. coconut, flaked
1 c. pecans, chopped

Directions:

1. Lightly butter an 8-inch square baking pan.
2. In 3-quart heavy saucepan, combine sugar, milk, corn syrup, and salt.
3. Cook over medium heat; stirring until sugar dissolves.
4. If sugar crystals form on sides of saucepan, wipe off.
5. Cook at low boil to softball stage (236 degrees F.).
6. Add butter without stirring.
7. Cool to lukewarm (110 degrees F.).
8. Add vanilla.
9. Beat until candy loses its gloss and starts to thicken.
10. Stir in coconut.
11. Pour at once into prepared pan.
12. While warm, mark into 36 pieces.
13. Cool until firm, and then cut.

Yields: 2 lbs.

Coconut Sweets

This makes an interesting, sweet coconut treat with the added protein of Ricotta cheese and milk powder.

Ingredients:

6 oz. fat-free Ricotta cheese
½ c. nonfat dry milk powder
10 Tbs. sugar
⅔ c. coconut, unsweetened, shredded, dried

Directions:

1. Grease a 9 x 13-inch square pan; set aside.
2. In large, nonstick frying pan, heat cheese over medium heat.
3. Add milk powder, mixing thoroughly.
4. Cook 12 to 15 minutes until most of the liquid is evaporated, stirring frequently.
5. Add sugar and stir; the mixture will become liquid again.
6. Cook 5 minutes, stirring occasionally.
7. Add coconut, mix thoroughly.
8. Continue to cook, 3 to 5 minutes; mixture should be quite thick.
9. Pour into prepared pan.
10. Press down with spatula.
11. Cut into 1 x 1-inch diamond shapes.
12. The mixture sets as it cools.
13. Cool completely before removing from the pan.

Yields: 24 servings.

Did You Know? . . . .

Did you know that copra (dried meat) is used to make coconut oil and is used in soaps, cosmetics, and hair oil?

Toasted Coconut Marshmallows

If you have never tried making homemade marshmallows you are in for a treat. Toasted coconut adds great flavor.

Ingredients:

7 oz. coconut, sweetened, shredded, toasted
powdered sugar

Ingredients for marshmallows:

3 pkg. unflavored gelatin
1½ c. sugar
1 c. light corn syrup
¼ tsp. kosher salt
1 Tbs. vanilla extract

Directions for marshmallows:

1. In small bowl, combine gelatin and ½ cup of cold water.
2. Allow to sit while making syrup.
3. In small saucepan, combine sugar, corn syrup, salt, and ½ cup water.
4. Cook over medium heat until sugar dissolves.
5. Raise heat to high; cook until the syrup reaches 240 degrees F. on candy thermometer.
6. Remove from heat.
7. With mixer on slow, slowly pour syrup into dissolved gelatin.
8. Put mixer on high; whip 15 minutes, until mixture is very thick.
9. Add vanilla and mix thoroughly.
10. Sprinkle half the toasted coconut in an 8 x 12-inch nonmetal pan.
11. Add batter and smooth top of mixture with damp hands.

12. Sprinkle on remaining toasted coconut.
13. Allow to dry uncovered at room temperature overnight.
14. Remove marshmallows from pan.
15. Cut into squares.
16. Roll sides of each piece carefully in sugar.
17. Store uncovered at room temperature.

Coconut Candies

Many children seem to enjoy peanut butter treats of all kinds. These are great with coconut.

Ingredients:

½ c. almonds
1½ c. coconut oil
¾ c. coconut flakes, dried
5 Tbs. cocoa powder
¼ c. honey
3 Tbs. butter
½ c. peanut butter

Directions:

1. In blender or food processor, grind almonds.
2. Add oil, coconut, cocoa powder, honey, butter, and peanut butter.
3. Process until well blended.
4. Pour into a 9 x 13-inch pan.
5. Note: You can also pour this into ice cube trays or cut into squares and refrigerate. They also freeze well.
6. These are a great treat to keep in the freezer as they make a handy snack or lunchbox treat.

Coconut Toffee

Try this change of pace with coconut-style toffee.

Ingredients:

1 c. thick coconut milk
1 c. milk
2 c. sugar
1 tsp. butter
few drops of vanilla extract
pinch of salt
few drops of food coloring (optional)

Directions:

1. In heavy saucepan, mix both milks together.
2. Heat on medium, bring to a boil.
3. Reduce heat; add sugar and salt, stirring constantly.
4. When mixture thickens, test by dropping a little in cold water. If it forms a soft ball, remove from heat.
5. Add butter and vanilla.
6. Return to heat, cook until it thickens.
7. Pour immediately onto a greased baking sheet.
8. Cut into squares.
9. Note: Food coloring may be added.

Yields: 25 pieces.

Coconut Haystacks

Coconut makes excellent candy. These are shaped in the form of a haystack. Enjoy!

Ingredients:

¾ c. evaporated milk
¾ c. brown sugar

6 Tbs. corn syrup
3 Tbs. butter
3 c. coconut, shredded

Directions:

1. In heavy 2-quart saucepan, add milk, sugar, and syrup.
2. Over low heat, cook and stir, until sugar dissolves.
3. Over medium heat, bring to boil, stirring often, until candy reaches soft ball stage – correct temperature is important.
4. Remove from heat.
5. Stir in coconut; mix well.
6. Shape into cones 1½-inches high.
7. Place on wax paper to cool.

Coconut Ice

Our family loves to make homemade ice cream. This is a very easy recipe with no cooking involved. It is great for children to make!

Ingredients:

3 c. powdered sugar
3½ c. coconut
1 can condensed milk
1½ tsp. vanilla extract
pink food coloring

Directions:

1. In medium bowl, mix all ingredients together.
2. Separate mixture into half.
3. Add a drop or two of coloring to one half.
4. Press white mixture into a pan; smooth until flat.
5. Spread pink mixture over top, smoothing also.
6. Refrigerate; cut into 2½-inch squares.

Coconut Potato Candy

This recipe is easy to make and is a unique way to use up leftover mashed potatoes.

Ingredients:

¾ c. mashed potatoes, cold
4 c. powdered sugar
4½ c. coconut, chopped or in flakes
1½ tsp. vanilla extract
¼ tsp. salt
4 squares baking chocolate, sweet or semi-sweet

Directions:

1. In large bowl, mix potatoes and sugar.
2. Stir in coconut, vanilla and salt; blend well.
3. Press into one large or two small pans, so candy will be ½-inch thick.
4. Melt chocolate over hot water, do not let water boil.
5. Pour chocolate on top of candy.
6. Cool and cut in squares.

Coconut Truffles

These coconut truffles are easy to make. Enjoy!

Ingredients:

1 pkg. white chocolate chips (8 oz.)
½ c. coconut cream plus 1 Tbs.
⅔ c. cup coconut, moist
1 Tbs. coconut liqueur
2 tsp. lime rind, grated
½ c. cocoa

Directions:

1. In small pan, add chocolate and cream.

2. Cook, over low heat, stirring constantly, 3 to 5 minutes until combined.
3. Remove from heat.
4. Stir in coconut, liqueur, and lime rind.
5. Transfer to a bowl, cover and refrigerate 1 hour.
6. Line a tray with wax paper.
7. Roll by 2 teaspoonfuls into a ball and place on a lined tray.
8. Refrigerate 30 minutes, or until firm.
9. In bowl, add cocoa.
10. Roll each ball in cocoa to coat.
11. Transfer to serving platter.

Coconut Chocolate Candy

Chocolate covered coconut candy is excellent. Be sure to make extra for company.

Ingredients:

- 1 c. sugar
- 1½ c. light corn syrup
- ½ c. water
- 1 pkg. coconut, flaked (14 oz.)
- ½ tsp. almond or vanilla extract
- 12 oz. chocolate chips, melted

Directions:

1. In large saucepan, combine sugar, syrup, and water.
2. Cook, over medium-low heat, stirring constantly, just until sugar is dissolved.
3. Cook without stirring to 236 degrees F. (soft ball stage).
4. Remove from heat.
5. Stir in coconut and flavoring; cool.
6. Shape into 1-inch balls.
7. Chill 1 hour.
8. Dip in melted chocolate if desired.
9. Note: You can also form in cone shapes and dip one end into chocolate.

Granola

It is always great to have some homemade granola on hand for breakfast. It is also great to take hiking or for lunch snacks.

Ingredients for granola:

3 c. rolled oats
¾ c. slivered almonds
1 c. walnuts, chopped
¼ c. sunflower seeds, shelled
½ c. coconut, unsweetened
½ c. raw wheat germ
¾ c. raisins, add after baking

Ingredients for syrup:

¼ c. maple syrup
⅓ c. water
¼ c. honey

Directions for granola:

1. Preheat oven to 225 degrees F.
2. In large bowl, mix all ingredients.

Directions for syrup:

1. Preheat oven to 300 degrees F.
2. In small bowl, mix syrup ingredients together.
3. Pour over dry mixture and stir.
4. Place in a 9 x13-inch baking pan.
5. Bake for 2 hours, stirring every 15 to 20 minutes.
6. Remove from oven, add raisins.
7. Allow to cool.
8. Store in airtight containers or zippered bags.

Coconut Delights Cookbook
A Collection of Coconut Recipes
Cookbook Delights Series – Book 4

Cookies

Table of Contents

Page

Cashew Coconut Date Cookies

My dad always liked date cookies. These are a variation of the usual date cookies but are delicious.

Ingredients:

1½ c. ground oats or oat flour
½ tsp. baking powder
½ tsp. baking soda
¾ tsp. sea salt
½ tsp. ground cardamom
½ c. maple syrup
⅔ c. butter
1 c. coconut, unsweetened, grated
¾ c. dates, finely chopped
1 c. flour
1 c. cashew butter, unsalted
3 Tbs. water
1 tsp. vanilla extract
1 c. granulated cane juice
40 cashew halves

Directions:

1. Preheat oven to 350 degrees F.
2. Line baking sheets with parchment paper.
3. In small bowl, combine oats, flour, baking powder, soda, salt, and cardamom; set aside.
4. In large mixing bowl, beat cashew butter and water.
5. Add syrup and vanilla.
6. Beat until well blended.
7. Add cane juice and butter; mix well.
8. Add dry ingredients.
9. Beat on low speed until blended well.
10. Stir in coconut and dates.
11. Drop dough by tablespoons or scoop.

12. Press down lightly with fingers and nestle one cashew half into each piece.
13. Bake 10 to 12 minutes until light brown.
14. Cool on wire rack.
15. Store in airtight container.

Coconut Cherry Bars

These cherry bars are enjoyed by all the maraschino cherry and coconut enthusiasts.

Ingredients:

1 c. flour
½ c. butter, softened
3 Tbs. powdered sugar
2 eggs
1 c. sugar
¼ c. flour
1 tsp. vanilla extract
½ tsp. baking powder
¼ tsp. salt
¾ c. nuts, chopped
¾ c. coconut, flaked
¾ c. maraschino cherries, drained, chopped

Directions:

1. Preheat oven to 350 degrees F.
2. Lightly grease an 8 or 9-inch square baking pan.
3. In small bowl, mix first 3 ingredients; press into pan.
4. Bake 10 minutes.
5. In medium bowl, beat eggs.
6. Stir in remaining ingredients; spread over bottom layer.
7. Bake 25 to 30 minutes, until golden brown.
8. Cool; cut into bars.

Chewy Coconut Macadamia Bars

These delicious bars may be made ahead of time. Wrapping well, they can be kept chilled for 4 days or frozen for 3 weeks. Serve bars chilled or at room temperature.

Ingredients for shortbread layer:

- 2 c. flour
- 1 c. butter, unsalted, cut into bits
- ⅔ c. powdered sugar
- ¼ tsp. salt

Ingredients for topping:

- ¼ c. butter, unsalted
- ½ c. brown sugar, firmly packed
- ¾ c. canned coconut cream
- ¼ c. heavy cream
- 2 Tbs. fresh lemon juice
- 1 pkg. coconut, sweetened, flaked coconut (7 oz.)
- 1 jar macadamia nuts, large pieces halved (7 oz.)

Directions for shortbread:

1. Preheat oven to 350 degrees F.
2. Butter a 9 x 13-inch pan.
3. Make shortbread layer.
4. In medium bowl, add shortbread ingredients.
5. Pat dough evenly onto bottom of prepared pan.
6. Bake 20 minutes, until golden.

Directions for topping:

1. In large saucepan, over low heat, melt butter.
2. Remove from heat.
3. Whisk in brown sugar until dissolved.

4. Add coconut cream, heavy cream, and lemon juice; whisking until well combined.
5. Stir in coconut and nuts.
6. Pour topping over shortbread.
7. Reduce oven temperature to 325 degrees F.
8. Bake until top is golden and center is bubbling.
9. Cool on a rack and cut into bars.

Yields: 40 3 x 1-inch bars.

Chocolate Coconut Cookies (No bake)

These are easy to make and children and adults all seem to enjoy them, as they disappear quickly.

Ingredients:

- 1¾ c. sugar
- 6 Tbs. cocoa powder, unsweetened
- ¾ c. evaporated milk
- ½ c. butter
- ½ tsp. vanilla extract
- 1¾ c. coconut, sweetened, shredded
- 3 c. rolled oats

Directions:

1. In medium saucepan, combine sugar, cocoa, evaporated milk, and butter.
2. Over medium heat, cook and stir, until mixture comes to a boil.
3. Boil 1 minute, stirring constantly.
4. Remove from heat; stir in vanilla.
5. Add rolled oats and coconut; mix well.
6. Drop by tablespoons onto a baking sheet lined with wax paper.
7. Refrigerate until firm.

Chocolate Crusted Coconut Bars

These are great bars with a combination of chocolate and delicious coconut.

Ingredients for crust:

- 1½ c. flour
- ¼ c. sugar
- 2 Tbs. cocoa powder, unsweetened
- 8 Tbs. butter, unsalted, room temperature

Ingredients for macaroons:

- 4 c. coconut, unsweetened, flakes
- 8 lg. egg whites
- 2 c. powdered sugar
- ½ c. sweetened coconut cream
- 2 oz. melted bittersweet chocolate, for decorating

Directions for crust:

1. Move oven rack to middle and preheat oven to 375 degrees F.
2. Lightly grease a 9 x 13-inch baking pan.
3. Put flour, sugar and cocoa in a bowl; add butter bit by bit, thoroughly blending.
4. Between 2 sheets of parchment paper, roll dough into a 9 x 13-inch rectangle.
5. Peel off 1 sheet of parchment and fit the dough into the bottom of prepared pan.
6. Peel off other sheet of parchment.
7. With a fork, prick holes in the dough.
8. Refrigerate while making the macaroons.

Directions for macaroons:

1. In medium bowl, mix coconut, egg whites, sugar, and cream together.
2. Lightly press mixture over crust.

3. Bake 15 minutes until golden brown.
4. Set on rack to cool.
5. Cut into bars 3 x 3-inches.
6. Drizzle with melted chocolate.

Coconut Almond Crispy Treats

Melted almond paste and marshmallows team up with crispy cereal and coconut that will delight kids and adults alike.

Ingredients:

1 c. coconut, sweetened, flakes
½ c. almonds, chopped or slivered
4 Tbs. butter
1 box almond paste, grated
1 bag mini marshmallows (10.5 oz.)
6 c. toasted rice cereal

Directions:

1. Spray with oil, or butter a 7 x 11-inch baking dish.
2. In large bowl, measure cereal and set aside.
3. In large nonstick skillet, add coconut and almonds.
4. Over medium heat, dry roast, stirring frequently, until golden brown.
5. Immediately pour into bowl with cereal.
6. In same skillet, melt butter over low heat.
7. Add the grated almond paste and marshmallows.
8. Stirring constantly, cook until all ingredients are melted and smooth.
9. Immediately pour almond-marshmallow mixture into bowl of cereal.
10. With buttered spoon mix thoroughly.
11. Press into prepared pan.
12. Cool until firm enough to cut.
13. Cut into 1½-inch squares

Yields: 32 squares.

Coconut Chocolate Orange Macaroons

If you like macaroons, you have to try this recipe.

Ingredients:

½ c. almonds, slivered, toasted
¾ c. coconut, sweetened, flakes
⅓ c. cocoa powder
1½ c. semi-sweet chocolate chips
2 tsp. orange peel, freshly grated
1 c. sugar
4 egg whites
2 Tbs. butter

Directions for almonds:

1. Preheat oven to 350 degrees F.
2. Spread almonds in a thin layer in shallow baking pan.
3. Bake 8 to 10 minutes, stirring occasionally, until light golden brown.
4. Cool completely.

Directions for cookies:

1. Reduce oven temperature to 325 degrees F.
2. Grind toasted almonds in food processor.
3. Line baking sheet with parchment paper or lightly spray with nonstick cooking spray.
4. Stir together ground almonds, coconut, and orange peel.
5. Stir together sugar and cocoa.
6. In mixer bowl, beat egg whites on medium-high until soft peaks form.
7. Gradually add sugar mixture, 2 tablespoons at a time, beating on high speed of mixer until stiff peaks form.
8. Gently fold almond mixture in egg white mixture.
9. Drop by slightly rounded teaspoons onto prepared baking sheet.

10. Bake 18 to 20 minutes, until firm.
11. Remove from baking sheet to wire rack.
12. Cool completely.
13. Combine chocolate chips and butter in a medium microwave-safe bowl.
14. Microwave on high, 1 minute; stir.
15. If necessary, microwave on high an additional 15 seconds at a time, stirring after each heating, until chips are melted when stirred.
16. Dip half of each cookie in chocolate mixture.
17. Place on wire racks over wax paper until chocolate is set.
18. Store cookies, loosely covered in containers, at room temperature.

Coconut Shortbread

My husband loves shortbread. This coconut shortbread is great served with your favorite tea.

Ingredients:

3 oz. butter
2 oz. coconut, dried
3 oz. flour
2 oz. fine tea sugar
2 Tbs. rice flour or fine semolina

Directions:

1. Preheat oven to 375 degrees F.
2. In small bowl, sift flour.
3. In medium bowl, cream butter and sugar.
4. Add flour, rice flour, and coconut; mix well.
5. Knead dough thoroughly.
6. Press dough onto ungreased baking sheet, forming a neat round.
7. Prick the round all over and mark into sections.
8. Bake 35 to 40 minutes.
9. Cool.

Coconut Cookie Bars

These are very easy to make for a quick coconut treat.

Ingredients:

½ c. butter
1½ c. graham cracker crumbs
1¾ c. coconut, flaked
1 can sweetened condensed milk (14 oz.)
1 pkg. semi-sweet chocolate chips (6 oz.)

Directions:

1. Preheat oven to 350 degrees F.
2. Melt butter in a 9 x 13-inch baking pan.
3. Sprinkle crumbs evenly over melted butter.
4. Pour milk evenly over crumbs.
5. Top evenly with remaining ingredients.
6. Press down firmly.
7. Bake 25 to 30 minutes, until light brown.
8. Cool before cutting.

Coconut Cookies

These are great coconut cookies to enjoy!

Ingredients:

½ tsp. salt
¼ c. evaporated milk
¼ c. water
1 c. coconut
3 c. flour
3 tsp. baking powder
1 tsp. vanilla extract
4 Tbs. butter
1 c. sugar
2 eggs, well beaten

Directions:

1. Preheat oven to 425 degrees F.
2. In large bowl, cream butter and sugar.
3. Add eggs; mix thoroughly.
4. Add vanilla and coconut, stir to blend.
5. In medium bowl, sift dry ingredients.
6. Add alternately with milk and water to first mixture.
7. Mix thoroughly.
8. Chill several hours.
9. Turn out onto lightly floured board.
10. Roll dough out ½-inch thick; cut with floured cutter.
11. Place on well-oiled baking sheet.
12. Bake 10 minutes.

Coconut Cookies (No bake)

These cookies are quick, easy, and delicious. Make extra because they will disappear quickly.

Ingredients:

2 c. sugar
½ c. butter
¾ c. whipping cream
2 tsp. vanilla extract
3 c. rolled oats
1¼ c. coconut
6 Tbs. cocoa

Directions:

1. In small bowl, combine oats, coconut, and cocoa.
2. In large saucepan, over medium heat, combine sugar, butter, whipping cream, and vanilla.
3. Let mixture come to foamy boil; remove from heat.
4. Add oats mixture to hot mixture, stir well.
5. Cool slightly; drop by teaspoonfuls onto wax paper.

Coconut Date Cookies (No bake)

These cookies are easy to make and taste great!

Ingredients:

½ c. butter
¾ c. pecans, chopped
½ c. maraschino cherries, drained, chopped
½ tsp. salt
1½ c. coconut, flaked
1 c. sugar
1 tsp. vanilla extract
2 c. cornflakes, crushed
2 Tbs. milk
1 egg, slightly beaten
8 oz. dates, chopped

Directions:

1. In 10-inch skillet, over medium heat, melt butter.
2. Stir in sugar and dates.
3. Remove from heat.
4. Stir in egg, salt, milk, and vanilla.
5. Cook, stirring occasionally, until mixture comes to full boil.
6. Boil, stirring constantly, 1 minute.
7. Remove from heat.
8. Stir in pecans, cornflakes, and cherries.
9. Shape into 1-inch balls.
10. Roll in coconut; place on wax paper.
11. Cover.
12. Refrigerate 1 hour until set.
13. Store in an airtight container.

Yields: 48 cookies.

Coconut Jam Tarts

Raspberry jam is great with coconut. Try these coconut jam tarts.

Ingredients for tart:

- 1 c. butter, softened
- 1 c. sugar
- 2 eggs, beaten
- 2 c. flour, sifted
- ½ c. jam, any flavor

Ingredients for topping:

- ¼ c. butter, softened
- ½ c. sugar
- 1 egg
- 1½ c. coconut, dessicated

Directions for tart:

1. Preheat oven to 350 degrees F.
2. In medium bowl, cream butter and sugar.
3. Stir in beaten eggs.
4. Mix in sifted flour with a fork.
5. When dough holds together and are well combined, shape into a ball.
6. Refrigerate 30 minutes.
7. Roll out to fit an 8-inch pie dish.
8. Fill with jam.

Directions for topping:

1. In small bowl, cream butter and sugar; beat in egg.
2. When combined, mix in coconut.
3. Spread this topping over the jam.
4. Bake 40 to 45 minutes.

Coconut Lemon Lassies

Our family loves lassies, and these add a nice variety to our collection of recipes.

Ingredients for cookies:

- 2½ c. flour, sifted
- 1 tsp. ground cinnamon
- ½ tsp. baking soda
- ¼ tsp. salt
- 1 c. sugar
- ½ c. butter, softened
- ¼ c. molasses
- 1 lg. egg

Ingredients for filling:

- ½ c. sugar
- ¼ c. lemon juice
- ⅛ tsp. salt
- 1 c. coconut, flaked
- 2 lg. eggs, slightly beaten
- 1 Tbs. lemon rind, grated
- 1 Tbs. butter

Directions for cookies:

1. Preheat oven to 350 degrees F.
2. Lightly grease baking sheets.
3. In bowl, sift together flour, cinnamon, baking soda, and salt.
4. In large bowl, beat sugar and butter with electric mixer until fluffy.
5. Beat in molasses and egg.
6. On low speed, gradually beat in flour mixture.
7. Chill 3 hours or overnight.

Directions for filling:

1. In small saucepan, cook and stir all ingredients except coconut; remove from heat.
2. Stir in coconut; cool.
3. Divide dough into four equal parts.
4. Roll one part at a time into a 15 x 2½-inch strip.
5. Spread ¼ of the filling on ½ of each strip, spreading to ¼-inch to edges.
6. Fold dough over filling; seal edges.
7. Cut into bars.
8. Place on baking sheets.
9. Bake 12 to 15 minutes.
10. Cool on wire racks.

Fruit Shaped Party Cookies

These cookies taste great and are fun to decorate.

Ingredients:

1½ pkg. fruit flavored gelatin
⅓ c. blanched almonds
½ tsp. almond extract (optional)
2 Tbs. sugar
1 can sweetened condensed milk
1 lb. macaroon coconut

Directions:

1. In medium bowl, mix ingredients together.
2. Chill well.
3. Shape a small amount into a desired fruit shape.
4. Roll in colored sugar.
5. Place a marzipan leaf in the stem end of each one.
6. Note: You may combine ½ box fruit-flavored gelatin with 4 tablespoons of sugar and a little food coloring for the colored sugar.

Orange Oatmeal Cookies

My dad always loved oatmeal cookies, and these are flavorful.

Ingredients:

½ c. butter plus ¼ tsp. for the pan
1 egg
¼ c. banana, ripe, mashed (½ banana)
1 tsp. vanilla extract
½ c. brown sugar, packed
1½ tsp. orange zest, grated or minced
1 c. whole wheat pastry flour
1 tsp. baking powder
1½ c. rolled oats
¾ c. coconut, unsweetened, shredded
1¼ c. walnuts, chopped
½ c. golden raisins or ½ c. chocolate chips

Directions:

1. Preheat oven to 350 degrees.
2. In large bowl, beat butter and egg until well blended and smooth.
3. Gradually beat in banana, vanilla, and sugar.
4. Add orange zest and mix until blended.
5. In a large bowl, mix together flour, baking powder, oats, coconut, walnuts, and raisins or chocolate chips.
6. Slowly add dry ingredients to wet ingredients and mix thoroughly.
7. Spread ¼ teaspoon of butter on baking sheet.
8. Drop heaping tablespoons of dough 2 inches apart.
9. Press down lightly.
10. Bake 20 minutes, until light brown.
11. Cool on wire rack.

Coconut Delights Cookbook
A Collection of Coconut Recipes
Cookbook Delights Series – Book 4

Desserts

Table of Contents

Page

Coconut Caramel Flan

Serve with glazed orange peel. Delicious!

Ingredients:

1½ c. sugar
½ c. water
1¼ c. whipping cream
¾ c. cream of coconut, well stirred
¼ c. coconut, sweetened, flaked, toasted
¾ c. coconut, sweetened, flaked
1 tsp. almond extract
3 Tbs. glazed orange peel, if desired
3 eggs
3 egg yolks
24 pc. sliced almonds, toasted

Directions:

1. Preheat oven to 350 degrees F.
2. In heavy medium saucepan, over low heat, heat sugar and water stirring until sugar dissolves.
3. Increase heat; boil until caramelized, 7 to 10 minutes.
4. Immediately pour caramel into an 8-cup soufflé dish, tilting and swirling dish to coat bottom and sides.
5. In large heavy saucepan, scald cream and cream of coconut.
6. Using electric mixer, beat eggs and yolks until thick and light.
7. Gradually mix in scalded cream.
8. Gently fold in ¾ cup flaked coconut and almond extract.
9. Pour into caramel-lined dish.
10. Place dish in large baking pan.

11. Pour enough water into baking pan to come halfway up sides of soufflé dish.
12. Bake flan 55 minutes, until knife inserted in center comes out clean.
13. Cool to room temperature.
14. Cover and refrigerate until well chilled.
15. Just before serving invert flan onto rimmed platter.
16. Arrange almonds in 8 fleur-de-lis patterns around upper edge of flan, using 3 slices for each.
17. Sprinkle toasted coconut between almonds.
18. Top with glazed orange peel, if desired.
19. Note: This can be prepared up to 1 day ahead.

Yields: 6 servings.

Coconut and Date Dessert

This is definitely an easy dessert to make and is best served warm, right out of the oven.

Ingredients:

1 lb. pitted dates, finely chopped
2 c. coconut, unsweetened, shredded
2 eggs
2 c. milk
whipping cream

Directions:

1. Preheat oven to 325 degrees F.
2. Lightly grease a 1-quart baking dish.
3. Sprinkle alternate layers of chopped dates and shredded coconut in prepared pan.
4. In small bowl, beat eggs with milk; pour over top.
5. Bake 45 minutes, until top brown.
6. Serve warm with cream.

Coconut Crème Brule

My family loves crème brule, and this is an excellent version.

Ingredients:

⅓ c. sugar
1 tsp. coconut rum
1 c. heavy cream
1 c. coconut milk
8 egg yolks

Directions:

1. Preheat oven to 325 degrees F.
2. In small saucepan, combine cream and milk.
3. Bring to a boil.
4. In small bowl, combine egg yolks, sugar, and coconut rum.
5. Once milk mixture has come to a boil, remove from heat.
6. Stir a small amount of the milk mixture (2 to 3 tablespoons) into the eggs yolks, whisking vigorously.
7. Slowly add remaining egg yolks back into hot milk mixture, whisking continuously.
8. Pour into individual glass serving dishes or custard cups.
9. With rack in center of oven, place cups on deep baking sheet filled with hot water.
10. Bake 15 to 20 minutes, until center is nearly set.
11. Remove from oven; cool on wire rack.
12. Sprinkle heavily with a layer of sugar.
13. Using a small torch, apply heat directly to surface of custard until a browned sugar layer forms.
14. Note: You may also place under broiler until light brown.

Coconut Gin Sherbet

This sherbet is refreshing served after an evening meal.

Ingredients:

½ lb. fresh coconut, grated
2¼ c. water
1¼ c. sugar
1 Tbs. gin

Directions:

1. In medium saucepan, simmer coconut in water for 20 minutes.
2. Add sugar, simmer 5 more minutes.
3. Cool until room temperature.
4. Strain, add gin.
5. Pour into ice cream maker.
6. Freeze until stiff.

Quick Coconut Sorbet

Try this quick and easy sorbet recipe.

Ingredients:

1 c. coconut cream
1 c. cold water
few drops almond extract

Directions:

1. In pan of hot water, warm opened can of coconut cream.
2. Pour into bowl, whisk until smooth.
3. Whisk in water; stir in extract.
4. Process in an ice cream maker; freeze until firm.

Coconut Custard

This makes a creamy custard dessert for all the coconut lovers.

Ingredients for caramel:

1 c. sugar
½ c. water

Ingredients for custard:

6 eggs
1 can condensed milk
1 can evaporated milk
½ tsp. vanilla extract
¾ c. coconut

Directions for caramel:

1. In small saucepan, mix water and sugar.
2. Boil until golden brown.
3. In an oven safe mold, cover bottom with caramel.

Directions for custard:

1. Preheat oven to 350 degrees F.
2. In blender, mix eggs and both milks for a few seconds.
3. Add coconut; mix for 25 seconds.
4. Pour mixture into mold with caramel.
5. Place mold inside a bigger mold half filled with water; place in oven.
6. Bake 1 hour and 15 minutes.
7. The custard is done when inserted toothpick comes out clean.
8. Cool and refrigerate.
9. When ready to serve turn out on a platter.

Banana Pineapple Coconut Icebox Dessert

This is a refreshing icebox dessert that your family and friends will ask you to make repeatedly. It can also be made ahead of company arriving.

Ingredients:

1 pkg. lg. marshmallows (16 oz.)
2 cans pineapple, crushed (15 oz.)
4 bananas
½ c. coconut, shredded
1 c. heavy whipping cream, whipped
2 c. graham or vanilla cracker crumbs

Directions:

1. Line a 9 x 13-inch baking pan with cracker crumbs.
2. Reserve a few crumbs as garnish for the top.
3. Drain pineapple, reserving juice.
4. In medium saucepan, over low heat, add pineapple juice.
5. Add marshmallows and melt.
6. Set aside to cool.
7. Cut up bananas into chunks.
8. Mix bananas, coconut, and pineapple into the cooled marshmallow mixture.
9. In large bowl, add whipped cream.
10. Fold in fruit mixture.
11. Spoon batter into crumb lined pan.
12. Sprinkle reserved crumbs on top.
13. Chill well before serving.
14. Cut into squares.

Yields: 14 to 18 servings.

Coconut Mint Sorbet

Sorbet is light and refreshing after a hearty meal. The coconut and peppermint adds great flavor.

Ingredients:

¾ c. sugar
2 bags peppermint herbal tea
1 lg. coconut without cracks and containing liquid

Directions:

1. Preheat oven to 400 degrees F.
2. In a saucepan, bring 2 cups water to a boil with sugar, stirring until sugar is dissolved.
3. Add tea bags; steep 15 minutes and discard tea bags.
4. With an ice pick or skewer, test the three eyes of the coconut to find the weakest one and pierce it.
5. Drain liquid and reserve.
6. Bake coconut for 15 minutes; break with a hammer.
7. Remove flesh from shell.
8. Cut coconut meat into small pieces.
9. Put small batches of coconut meat in blender and grind transferring to a bowl as done.
10. Return the ground coconut to the blender.
11. Strain the reserved coconut liquid through a fine sieve into a large measuring bowl; add enough water to make 2 cups liquid.
12. In a saucepan, bring liquid to a boil.
13. Add to the ground coconut and blend for 1 minute.
14. Cool; strain through a large sieve into a bowl lined with a double thickness of cheesecloth pressing hard on the solids and squeeze to extract as much milk as possible.
15. Stir tea into the coconut milk.
16. Chill covered until mixture is cold.

17. Put into freezer containers and freeze; keep frozen.
18. Sorbet may be made 5 days in advance.
19. Let sorbet stand in the refrigerator for 30 minutes, or microwave at 30 percent power for 1 minute to soften slightly before serving.

Coconut Flan

My husband loves flans and the coconut milk gives this version a great flavor.

Ingredients:

2 c. thick coconut milk
6 egg yolks
4 egg whites
1 c. sugar, refined
1 c. brown sugar

Directions:

1. In small saucepan, over medium heat, dissolve brown sugar in ¼ cup water.
2. Cook until sugar browns or caramelizes.
3. Line a mold evenly with ¾ of caramelized syrup; set aside.
4. Stir milk into remaining ¼ of caramelized syrup.
5. Place over low heat; stir continuously until all caramel is dissolved.
6. Mix egg yolks, and slightly beaten whites, beat lightly.
7. Add sugar and lemon rind.
8. Add coconut milk with caramelized syrup; mix well.
9. Strain through cheese cloth.
10. Pour into mold.
11. Cook slowly in a pan with hot water; without allowing water to boil.

Yin Yang Papaya Dessert

This is a refreshing summer dessert with a nice contrast of textures and flavors.

Ingredients:

1 c. unsweetened coconut milk, well stirred
¾ c. milk
3 Tbs. sugar
2 Tbs. quick-cooking tapioca
1 tsp. vanilla extract
2 med. ripe papayas
fresh mint sprigs, for garnish

Directions:

1. In medium saucepan, over high heat, combine both milks, sugar, and tapioca.
2. Reduce heat to low; simmer, stirring frequently until tapioca is slightly thickened, 5 minutes.
3. Stir in vanilla.
4. Let cool.
5. Cover and refrigerate, 1 hour.
6. Halve papaya and remove seeds.
7. With spoon, scoop out flesh.
8. Place in a blender or food processor.
9. Process papaya to a purée, total of 2 cups.
10. Cover and refrigerate until well chilled, at least 1 hour.
11. To serve, spoon tapioca into one side of each soup bowl.
12. Add papaya purée on other side, forming a yin-yang pattern.
13. Garnish with mint sprig.

Yields: 4 servings.

Pumpkin Coconut Dessert

This dessert is creamy and it has the sweet and spicy, earthy flavor of fresh baked pumpkin.

Ingredients:

1½ c. sugar
¼ tsp. salt
1 med. sugar pumpkin, 3.5 lb.
1 can coconut milk
1 tsp. ground cardamom
1 tsp. ground cinnamon

Directions:

1. Halve and seed pumpkin.
2. Cut each half into four equal pieces.
3. In large pan, steam pumpkin in 1 inch of lightly salted boiling water for 15 minutes.
4. Drain in colander; cool.
5. In same pan, mix milk, sugar, cardamom, cinnamon, and salt; bring to boil.
6. Turn down heat, simmer, stirring frequently until sugar is dissolved.
7. Slice or peel skin from pumpkin.
8. Cut into small bite-size squares.
9. Add pumpkin to milk mixture.
10. Simmer 10 minutes.
11. Cool to room temperature and refrigerate.
12. Serve warm or cold.
13. Top with whipped cream or serve with a side of warm gingerbread.
14. Note: Sugar pumpkins are not jack-o-lantern pumpkins. They are a sweeter, smaller variety grown for cooking.

Yields: 8 to 10 servings.

Coconut Pumpkin Pudding

Use your autumn pumpkins with coconut and you will have an attractive, delicious pudding.

Ingredients:

½ pumpkin, peeled, chopped
1 c. sugar
6 med. half-ripe coconuts with good flesh inside
salt, to taste

Directions:

1. Preheat oven to 350 degrees F.
2. Cut tops off coconuts.
3. Pour out coconut water; set aside.
4. Add sugar and salt to pumpkin.
5. Mix well with a small amount of coconut water.
6. Fill coconut shells with mixture.
7. Bake 1 hour.
8. Cool and serve with thick coconut cream

Homemade Coconut Ice Cream

Our family enjoys homemade ice cream and coconut adds great flavor.

Ingredients:

1 c. heavy cream
½ c. milk
½ c. coconut flakes, packed
1 can sweetened coconut cream (8 oz.)

Directions:

1. In blender, add milk and coconut cream.
2. Blend until smooth.
3. Carefully stir in cream and coconut flakes.
4. Transfer mixture to ice cream maker, follow the manufacturer's instructions.

Carrot Coconut Cheesecake Ball

This is a flavorful cheesecake dessert that makes an attractive presentation.

Ingredients:

1 can pineapple, crushed (8 oz.)
¼ c. coconut flakes
½ c. carrots, finely shredded
½ c. raisins
1 pkg. cream cheese (8 oz.)
1 pkg. light cream cheese (4 oz.)
⅓ c. sugar
½ tsp. ginger, powdered
½ tsp. nutmeg
1 tsp. cinnamon
1 tsp. vanilla extract
1 c. graham cracker crumbs

Directions:

1. Drain can of pineapple, soak up excess juice with paper towel. Set aside.
2. In small bowl, combine dry ingredients.
3. Add carrots and raisins.
4. In medium bowl, mix pineapple, coconut, vanilla, and cream cheeses.
5. Add dry ingredients.
6. Form into a ball.
7. Roll in graham cracker crumbs.

Did You Know? . . . .

Did you know that coconut shells are used for utensils, cups, bowls, bottles, lamps, buckles, and ornaments?

Coconut Cheesecake

Cheesecake is always a welcome finale to a meal, and this coconut cheesecake is delicious.

Ingredients for crust:

1½ c. graham cracker crumbs
¾ c. sugar
⅓ c. butter, unsalted, melted

Ingredients for filling:

1 c. whipping cream
3 c. coconut, sweetened, shredded, toasted
1 tsp. vanilla extract
1 Tbs. lemon juice, freshly squeezed
2 lb. cream cheese, room temperature
4 lg. eggs
1 can cream of coconut (15 oz.)
juice of 1 lemon

Directions:

1. Preheat oven to 325 degrees F.
2. Wrap outside of a 9-inch springform pan with 2¾-inch high sides with foil.
3. Mix graham cracker crumbs, coconut, lemon juice, and butter in small bowl.
4. Press mixture onto bottom and sides of pan.
5. In large bowl, beat cream cheese and sugar until well blended.
6. Add eggs one at a time, beating well after each addition.
7. Add cream of coconut, whipping cream, coconut, vanilla, and lemon juice; blend well.
8. Pour filling into crust.

9. Bake 1 hour and 25 minutes until puffed and set in center.
10. Cool completely on wire rack.
11. Cover and refrigerate overnight.
12. Cut around pan sides to loosen cake.
13. Remove pan sides.
14. Sprinkle remaining coconut around edge of cake.

Lime and Coconut Pudding

You will love this delicious lime and coconut pudding.

Ingredients:

½ c. sugar
2 Tbs. butter, soft
2 Tbs. flour
4 Tbs. coconut, shredded or desiccated
1 c. buttermilk or milk
2 eggs, separated
2 lg. limes, juice and zest
thick cream
powdered sugar

Directions:

1. Preheat oven to 350 degrees F.
2. Lightly grease an ovenproof dish.
3. In small bowl, beat sugar and butter.
4. Add lime juice, zest, flour, coconut, milk, and egg yolks.
5. Mix well.
6. In another small bowl, beat egg whites until stiff.
7. Fold into mixture.
8. Pour into prepared dish or cups.
9. Bake 25 to 30 minutes, until top is golden.
10. Serve warm with cream.
11. Sprinkle with powdered sugar.

Fried Banana Coconut Supreme

These rolls are a lot of work but well worth the trouble.

Ingredients:

- ½ c. powdered sugar
- ¼ c. sesame seeds
- ¼ c. coconut, flaked
- 1 egg yolk
- 2-3 c. canola oil
- 1 pkg. egg roll or spring roll wrappers, thawed
- 1 bunch bananas

Directions:

1. Cut banana in half on wrapper.
2. Slice each banana into ½-inch slices, but not through.
3. Wrap up like an envelope.
4. Use egg yolk to seal corners and flap.
5. In ungreased skillet, over low heat, add sesame seeds.
6. Stir constantly until golden brown.
7. Remove and cool.
8. Crush with rolling pin.
9. In small bowl, add sesame seeds, sugar, and coconut; mix well.
10. Heat oil to 365 degrees F.
11. Deep fry rolls, 2 or 3 at a time, turn once, until golden brown.
12. Drain on paper towels.
13. Spread mixture on top.
14. Serve hot.
15. Serve with vanilla ice cream or melt semi-sweet chocolate and drizzle on rolls and ice cream.

Coconut Delights Cookbook
A Collection of Coconut Recipes
Cookbook Delights Series – Book 4

Dressings, Sauces, and Condiments

Table of Contents

Page

Coconut Chutney

Serve this as a delicious dip for snacks.

Ingredients:

- 1 c. coconut, unsweetened, grated
- ½ c. onion, finely chopped
- 1 tsp. minced ginger
- 1 tsp. minced green chili, serrano or jalapeno pepper
- ¼ c. plain yogurt
- 1 tsp. fresh lemon juice
- ½ tsp. salt
- ½ c. water, as needed
- ½ tsp. mustard seeds
- 2 Tbs. olive oil
- 2 dried red peppers
- 8 curry leaves or 2 bay leaves

Directions:

1. In blender or food processor, grind coconut, onion, ginger, chili, yogurt, lemon juice, and salt with as little water as needed for a moderately thick consistency.
2. Set aside.
3. In covered pan, over medium heat, add oil, mustard seeds, peppers, and curry leaves.
4. Cook until seeds begin to pop.
5. Uncover; add ground coconut mixture, stir for 10 seconds.
6. Remove from heat.
7. Transfer to serving bowl.

Did You Know? . . . .

Did you know that copra meal (dried meat of the coconut) is used as animal feed for horses and cattle?

Coconut Dressing

This is a delicious dressing to serve with fruit kabobs. Serve cold with your favorite fresh, cold fruits on a skewer.

Ingredients:

1½ c. vanilla or coconut yogurt
1½ Tbs. coconut, flaked
1½ Tbs. orange marmalade

Directions:

1. In small bowl, combine all ingredients.
2. Stir well.

Coconut Fudge Sauce

This is a delicious fudge sauce to serve over ice cream.

Ingredients:

¾ c. evaporated milk
½ c. sugar
1½ tsp. butter
2 tsp. coconut extract
1 pkg. semi-sweet chocolate chips (6 oz.)

Directions:

1. In large saucepan, over medium heat, add milk, chocolate chips, and sugar.
2. Stir constantly until chocolate is melted and mixture begins to boil.
3. Remove from heat; stir in butter and extract.
4. Pour into jars; cover tightly.
5. Serve warm or cold over ice cream.
6. Note: Refrigerate no longer than 4 weeks.

Coconut Lime Dressing

This is a delicious dressing to serve with fresh fruit on a hot summer day.

Ingredients:

½ tsp. coconut extract
1 tsp. ginger purée
2 tsp. lime rind, grated
1 Tbs. plus 1 tsp. honey
½ c. yogurt
1 tsp. mint, chopped

Directions:

1. In small bowl, combine all ingredients.
2. Chill, one hour.
3. Serve over sliced fresh fruit.

Coconut Miracle Whip

Try this miracle whip.

Ingredients:

4 egg yolks
1 tsp. sea salt
2 tsp. dry mustard
6 Tbs. apple cider vinegar
2 c. safflower oil
1 c. coconut milk or oil
3 Tbs. cornstarch or flour
1 c. water
4 Tbs. honey
¼ c. apple cider vinegar
1 Tbs. sea salt
¼ c. lemon juice, freshly squeezed

Directions:

1. In medium bowl, using a mixer on high speed, blend yolks, salt, dry mustard, and 2 tablespoons vinegar.
2. Very slowly add safflower oil.
3. Mix thoroughly.
4. Gradually add remaining safflower oil, blending well after each addition.
5. Add remaining 4 tablespoons of vinegar and lemon juice.
6. In medium saucepan, add remaining ingredients while heating.
7. Cook to a smooth paste, stirring often.
8. Blend hot mixture slowly into original mixture continuing to blend with mixer on high speed.
9. Pour into a container.
10. Cool in refrigerator.
11. Will keep in refrigerator for 1 to 2 weeks.

Coconut Thousand Island Dressing

This is a twist on the usual Thousand Island dressing. It is delicious.

Ingredients:

½ c. coconut mayonnaise (see page 184)
¼ c. pickle relish
2 Tbs. ketchup
⅛ tsp. paprika
salt and pepper, to taste

Directions:

1. In small bowl, combine all ingredients.
2. Serve on your favorite salad.

Coconut Sauce for Fish

This makes a great and flavorful sauce for fish.

Ingredients:

2 mature coconuts, finely grated
4 cooked prawns or shrimp
¼ c. seawater

Directions:

1. Shell and remove heads of shrimps or prawns.
2. Wrap in muslin cloth.
3. Crush wrapped prawns or shrimp into coconut using a rock.
4. Occasionally dip in seawater and continue to crush until nothing is left inside the cloth.
5. Mix well.
6. Line a bowl with softened banana leaves.
7. Pack coconut mixture in bowl and cover with banana leaf.
8. Use coconut shells to hold down leaves.
9. Serve in shells as sauce.
10. Note: Chiles may be used for flavoring.

Coconut Curry Sauce

Our family loves curry, and this is a great sauce.

Ingredients:

1 Tbs. olive oil
1 med. onion, thinly sliced
4 garlic cloves, minced
4 tsp. curry powder
2 cans unsweetened coconut milk (14 oz. ea)
1 c. chicken stock or canned low-salt chicken broth

2 med. carrots, peeled, cut into matchstick-size pieces

Directions:

1. In heavy medium saucepan, heat oil over medium heat.
2. Add onion, garlic, and curry powder.
3. Sauté mixture, 4 minutes.
4. Mix in coconut milk and stock.
5. Bring to boil.
6. Reduce heat to medium-high; boil until liquid is reduced to thin sauce consistency, 50 minutes.
7. Add carrots.
8. Simmer until crisp-tender, about 3 minutes.
9. Remove from heat.
10. Serve with your favorite fish, chicken, or pork recipe.

Creamy Italian Coconut Dressing

Try this Italian Dressing using coconut oil.

Ingredients:

1 pt. sour cream
¼ c. coconut oil
½ tsp. each oregano, basil, tarragon, and pepper
¼ tsp. rosemary
3 Tbs. apple cider vinegar
1 tsp. garlic powder
pinch of stevia
sea salt, to taste

Directions:

1. In small bowl, combine all ingredients.
2. Note: This dressing is best made the day before.

Curried Chocolate Coconut Sauce

This sauce is delicious served over ice cream.

Ingredients:

6 oz. bittersweet or semi-sweet chocolate, chopped
2 Tbs. butter
¼ c. canned sweetened cream of coconut
2 Tbs. dark rum
2 tsp. curry powder
1 tsp. lime peel, finely grated

Directions:

1. In medium saucepan, over low heat, stir chocolate and butter until smooth.
2. Whisk in next 4 ingredients.
3. Stir sauce until just warm.
4. Note: This sauce can be made 1 day ahead. Cover and chill. Reheat to serve.

Marmalade Dipping Sauce

This sauce is especially delicious when served with coconut shrimp.

Ingredients:

½ c. orange marmalade
2 tsp. stone-ground mustard
1 tsp. prepared horseradish
dash of salt

Directions:

1. In small bowl, mix all ingredients together; cover.
2. Place in refrigerator 1 hour before serving.

Curry Coconut Dressing

This makes a great dipping sauce for vegetables or meat.

Ingredients:

1 c. unsweetened coconut milk
¼ c. lime juice
2 Tbs. jalapeno jelly, melted
2 tsp. fresh gingerroot, grated
1 tsp. curry powder
salt, to taste

Directions:

1. In large bowl, combine all ingredients.
2. Mix well.

Dried Coconut Chutney

This is a great chutney for dipping and is also easy to make.

Ingredients:

1 c. coconut, dried, grated
5 garlic cloves, chopped
1 tsp. cumin
2 tsp. chili powder
7 red chilies
½ lg. onion, chopped
salt, to taste

Directions:

1. In small bowl, combine all ingredients.
2. Grind into a mixture.
3. Serve with Chapati or your favorite bread.

Mayonnaise

My husband and I used to always make our own mayonnaise when we were first married. This recipe has a great flavor. Remember that eating raw eggs is not recommended, so do your own research before you do.

Ingredients:

2 egg yolks
2 Tbs. fresh lemon juice
½ tsp. Dijon mustard
⅛ tsp. paprika
⅛ tsp. sea salt
¼ c. olive oil
½ c. coconut oil

Directions:

1. In food processor or blender, combine all ingredients except coconut oil.
2. Slowly drizzle in coconut oil until emulsified into mayonnaise.

Pineapple Orange Coconut Dressing

Try this dressing on your favorite fruit or green salad. Also try this recipe with apple and orange concentrates.

Ingredients:

½ c. coconut milk
4 Tbs. each of pineapple and orange concentrate

Directions:

1. In small bowl or blender, mix both ingredients well.
2. Top salad with extra shredded coconut, orange, and pineapple bits.

Pina Colada Dipping Sauce

This delicious sauce goes perfectly with shrimp.

Ingredients:

1 c. pina colada mix
¼ c. water
2 Tbs. pineapple, crushed, drained
1 Tbs. and 1 tsp. coconut flakes, sweetened
3 Tbs. and 1 tsp. powdered sugar
1½ tsp. cornstarch
3 tsp. cold water

Directions:

1. In medium saucepan, combine pina colada mix, water, crushed pineapple, coconut, and powdered sugar.
2. Heat on medium-low until sauce begins to simmer, stirring frequently.
3. Simmer slowly, 10 to 12 minutes.
4. Mix cornstarch and water together.
5. Add to sauce and blend well stirring constantly.
6. Simmer 3 to 5 minutes longer; remove from heat.
7. Serve at room temperature.

Coconut Mustard Sauce

Serve this easy-to-make sauce with your favorite appetizers. It is simple and flavorful.

Ingredients:

½ c. coconut cream
½ c. Dijon mustard

Directions:

1. In small bowl, combine ingredients; mix well.

Spicy Pork Coconut Sauce

This great sauce is a combination of sweet, hot, and citrus tangy. It is a great combination with pork loin. Serve with rice, salad, and your favorite bread for a complete meal.

Ingredients:

1½ lb. pork loin, cut in pieces
⅔ c. coconut cream
½ c. fresh lime juice
1 tsp. lime zest, grated
6 Tbs. green onion and tops, minced
½ tsp. cayenne pepper
2 tsp. chili powder
½ tsp. salt
1 garlic clove, minced
1 jalapeno pepper, chopped
¼ c. cilantro, chopped
1 tsp. grated ginger or ½ tsp. ground
choose one: 2 tsp. curry powder, 2 Tbs. peanut butter or 1 Tbs. fish sauce

Directions:

1. In small bowl, whisk coconut cream and fresh lime juice until smooth.
2. Stir in lime zest, green onions, cayenne, chili powder, salt, garlic, jalapenos, cilantro, ginger, and your choice of curry powder, peanut butter or fish sauce.
3. Marinate pork up to 24 hours.
4. Cook or bake over moderate heat.
5. Note: Sauce may be made a day ahead.

Yields: 1½ cups.

Spicy Shrimp Coconut Sauce

This great sauce is a combination of sweet, hot, and citrus tangy. It is a great combination with shrimp. Serve with rice, salad, and your favorite bread for a complete meal.

Ingredients:

1½ lb. shrimp
⅔ c. coconut cream
½ c. fresh lime juice
1 tsp. lime zest, grated
6 Tbs. green onions and tops, minced
½ tsp. cayenne pepper
2 tsp. chili powder
½ tsp. salt
1 garlic clove, minced
1 jalapeno pepper, chopped
¼ c. cilantro, chopped
1 tsp. fresh ginger, grated or ½ tsp. ground
choose one: 2 tsp. curry powder, 2 Tbs. peanut butter or 1 Tbs. fish sauce

Directions:

1. In small bowl, whisk coconut cream and fresh lime juice until smooth.
2. Stir in lime zest, green onions, cayenne, chili powder, salt, garlic, jalapenos, cilantro, ginger, and your choice of curry powder, peanut butter or fish sauce.
3. Marinate shrimp up to 24 hours.
4. Cook or bake over moderate heat.
5. Note: Sauce may be made a day ahead.

Yields: 1½ cups.

Golden Coconut Fruit Dressing

This is a flavored dressing for fruit.

Ingredients:

4 tsp. cornstarch
⅓ c. pineapple juice
⅔ c. coconut cream
2 egg yolks, beaten
8 oz. pineapple yogurt
dash salt

Directions:

1. In small saucepan, dissolve cornstarch in pineapple juice.
2. Stir in coconut cream and salt.
3. Over medium heat, cook and stir until mixture comes to a boil.
4. Slowly stir ¼ cup hot cornstarch mixture into egg yolks; add to remaining cornstarch mixture in pan.
5. Cook and stir until thickened, 1 minute.
6. Cool; chill thoroughly.
7. Fold in yogurt.
8. Serve with fresh fruit.

Simple Tartar Sauce

This is very easy to make quickly for your favorite tartar sauce dip.

Ingredients:

½ c. coconut mayonnaise
1 Tbs. dill pickle relish

Directions:

1. In small bowl, combine both ingredients.
2. Serve with your favorite dish.

Coconut Delights Cookbook
A Collection of Coconut Recipes
Cookbook Delights Series – Book 4

Jams, Jellies, and Syrups

Table of Contents

Page

A Basic Guide for Canning Jams, Jellies, and Syrups

1. Wash jars in hot, soapy water inside and out with brush or soft cloth.
2. Run your finger around rim of each jar, discarding any with cracks or chips.
3. Rinse well in clean, clear, hot water, using tongs to avoid burns to hands or fingers.
4. Place upside down on clean cloth to drain well.
5. Place lids in boiling water for 2 minutes to sterilize and keep hot until placing on rim of jar.
6. Immediately prior to filling each jar, immerse in very hot water with tongs to heat jar (avoids breakage of jar with hot liquid).
7. Fill jar to within 1 inch of top of rim or to level recommended in recipe.
8. Wipe rim with clean damp cloth to remove any particles of food, and check again for any chips or cracks.
9. With tongs, place lid from hot bath directly onto rim of jar.
10. Using gloves, cloth, or holders, tighten lid firmly onto jar with ring or use single formed lid in place of ring to cover inner lid. Do not tighten down too hard as it may impede sealing.
11. Place on protected surface to cool, taking care to not disturb lid and ring. A slight indentation of lid will be apparent when sealed.
12. Leave overnight until thoroughly cooled.
13. When cooled, wipe jars with damp cloth and then label and date each.
14. Store upright on shelf in cool, dark place.

Coconut Butter

Try this easy, flavorful butter on your next honey bread, pancakes, or toast.

Ingredients:

1 c. butter
¼ c. cream of coconut

Directions:

1. In small mixing bowl, add butter and cream of coconut.
2. Mix with electric mixer until fluffy.
3. Serve with your favorite bread or rolls.

Coconut Syrup

Serve this syrup on your favorite ice cream or yogurt. You can also try it on pancakes, waffles, or French toast.

Ingredients:

1 can unsweetened coconut milk
1 c. coconut, shredded
¾ c. sugar

Directions:

1. In heavy saucepan, combine all ingredients.
2. Bring to a boil, reduce to simmer.
3. Cook 20 minutes, stirring occasionally.
4. Transfer to blender.
5. Purée until smooth.
6. Serve immediately.

Coconut Jam

Try this coconut jam. It is great served on English muffins or your favorite bread or toast.

Ingredients:

1 lb. coconut, unsweetened, flaked
2 Tbs. orange-blossom water or rosewater
2 c. sugar
1 Tbs. fresh lemon juice
2 c. unsalted pistachios
2 c. blanched almonds, chopped

Directions:

1. In bowl, sprinkle coconut with flower water and 4 cups of water, fluffing as you do so.
2. Cover and leave overnight.
3. In saucepan, combine sugar, lemon juice, and 2 cups water.
4. Simmer 5 minutes, stirring constantly to thicken.
5. Add soaked coconut, slowly bring to a boil again stirring constantly.
6. Remove from heat as soon as mixture boils, or it will harden and yellow.
7. Let cool.
8. Mix in nuts.
9. Pour into glass bowl.
10. Serve the following day or refrigerate in tightly sealed sterilized jars.
11. Note: Orange-blossom water and rosewater is available at gourmet and Middle East markets.

Did You Know? . . . .

Did you know that the midribs of the leaves can be collected and made into a bundle and used as a broom?

Coconut Egg Jam

This is an unusual jam but fun to make and actually very tasty. Be sure to try it and share with friends.

Ingredients:

1 can coconut milk
1½ c. sugar
10 eggs

Directions:

1. In mixing bowl, beat eggs at medium speed with mixer.
2. Add sugar; mix thoroughly at high speed.
3. Add coconut milk; mix well at high speed, making sure sugar has dissolved.
4. Pour into a pot and cook over low heat stirring constantly.
5. Note: The mixture will eventually change from cream egg-white to a red-brown color; this is due to the caramel forming from the sugar.
6. Continue stirring to prevent the jam from burning at the bottom, lowering heat if necessary.
7. Once there is no streaks left in the jam, remove from stove and cool.
8. Pour into sterilized jars; refrigerate.
9. Serve on bread or toast.

Did You Know?. . . .

Did you know that different sections of the leaves can be made into many different types of baskets and hats and many other creative items?

Coconut Spread

This is a simple recipe to enjoy on your favorite English muffin or toast.

Ingredients:

8 oz. cream cheese, softened
5 Tbs. apricot, peach or pineapple preserves
¾ c. coconut, flaked

Directions:

1. Combine cream cheese and preserves, mixing until well blended.
2. Add coconut; mix well.
3. Chill.

Apricot Coconut Spread

This spread is great served with pancakes, bread, or over porridge. It is even great as pie filling or pudding.

Ingredients:

1 c. dried apricots
½ c. hot water
⅓ c. coconut cream
1 Tbs. coconut spread, well blended
⅛ tsp. ground cloves
⅛ tsp. nutmeg

Directions:

1. Place apricots in water; let stand 30 minutes, until they soften.
2. Drain, reserve water.
3. In food processor, purée all ingredients including water.
4. Add a pinch of stevia if not sweet enough.

Coconut Delights Cookbook
A Collection of Coconut Recipes
Cookbook Delights Series – Book 4

Main Dishes

Table of Contents

Page

Coconut Shrimp with Honey Mustard Horseradish Sauce

Honey and mustard makes a great sauce for these flavorful coconut shrimp.

Ingredients for sauce:

1½ c. honey
½ c. Dijon mustard
½ c. horseradish
1 c. mayonnaise

Ingredients for shrimp:

2 c. flour
½ qt. fry batter mix (flour, cornmeal, salt, spices)
½ lb. coconut flakes
6 oz. coconut milk
65 med. shrimp, peeled, deveined
oil for frying

Directions for sauce:

1. In small bowl, mix all ingredients well.

Directions for shrimp:

1. Heat oil to 375 degrees F.
2. Dredge shrimp in flour to coat.
3. Mix fry batter and coconut milk together adding water, if necessary, to create cake batter consistency.
4. Dip shrimp in batter mixture.
5. Roll shrimp in coconut flakes to cover.
6. Deep fry to golden brown.
7. Serve with sauce.

Coconut Kabocha Stew

This is a flavorful and healthy vegetarian stew to try.

Ingredients:

3½ c. Kabocha squash, baked, peeled
4 lg. garlic cloves, minced
1 onion, diced
4 stalks celery, diced
2 c. water
1 Tbs. lemon juice
1 can coconut milk (14 oz.)
1 Tbs. miso (optional)
2 c. frozen corn
10 oz. frozen spinach
1 can garbanzo beans, rinsed, drained
cayenne pepper, to taste

Directions:

1. Preheat oven to 375 degrees F.
2. Rinse squash, pierce with a knife in several places.
3. Place on baking sheet.
4. Bake 40 minutes, until very soft.
5. In large pot or Dutch oven, bring garlic, onion, celery, and water to boil.
6. Simmer, 10 minutes.
7. Peel and measure 3½ cups squash.
8. Mash squash with fork or potato masher.
9. Leave some chunks about 1-inch square.
10. Measure and thaw frozen vegetables.
11. Add squash to vegetables, blending well.
12. Reduce heat to lowest setting.
13. Add lemon juice and milk.
14. Stir occasionally until thoroughly blended.
15. Stir in miso and cayenne if desired.
16. Stir in spinach, corn, and beans.
17. Serve hot with your favorite bread or buns.

Beef with Pineapple and Coconut Curry

Our family enjoys curry, and this is very easy to make and great with rice.

Ingredients:

1 c. yellow onions, coarsely chopped
3 Tbs. curry powder
2 Tbs. salt
1 tsp. black pepper
½ c. olive oil
½ fresh pineapple, trimmed, cut into 2-inch chunks
½ c. raisins
4 lb. boneless chuck, cut into 2-inch squares
2 c. rice
5 c. salted water

Ingredients for topping:

mango, peeled, cut into long strips
fresh coconut, cut into 1-inch squares

Directions:

1. In large skillet, add oil.
2. Over medium heat, sauté onions, curry powder, salt, and pepper until onions are soft.
3. Add boneless chuck.
4. Cover pan immediately.
5. Do not allow meat to brown.
6. Simmer slowly for 1 hour.
7. Add pineapple and raisins.
8. Place in serving bowl.
9. Pierce fresh coconut; drain water and save for future use.
10. Place mango strips and coconut squares on top.

Caribbean Coconut Shrimp

Caribbean flavors and coconut make a delicious shrimp.

Ingredients:

1 lb. lg. or jumbo shrimp, peeled, deveined
¾ c. flour
1 egg
½ Tbs. baking powder
½ c. beer
¼ c. flour, for coating
1 Tbs. salt
½ Tbs. ground pepper
½ Tbs. cayenne
1½ c. coconut, dried, grated
½ Tbs. paprika
1 Tbs. garlic powder
½ tsp. dried thyme
½ tsp. oregano

Directions:

1. Dip each shrimp into batter, roll in coating.
2. Deep fry.
3. Allow to drain on paper towel.
4. Serve with your favorite sauce or dip.

Did You Know? . . . .

Did you know that some people believe that wearing coconut jewelry often brings travelers good luck?

Did you know that a section of the breast-like section of the leaf can be used to make a fan?

Chicken Colombo and Creole Rice

This is a delicious and easy-to-make chicken dish that makes a unique change of pace from the usual chicken recipes. Try it for an enjoyable dinner, and serve with Creole Rice as a great accompaniment to the chicken (recipe below).

Ingredients for chicken:

- 3 Tbs. olive oil
- 2 Tbs. butter
- 2 lb. chicken, select a mix of your favorite pieces
- 4 Tbs. Colombo powder
- 2 c. chicken broth
- 2 c. onions, peeled, chopped
- 4 Tbs. freshly squeezed lime juice
- 2 Tbs. fresh thyme
- 3 Tbs. chives, chopped
- 3 Tbs. parsley, chopped
- 1 tsp. green or red pepper, chopped
- 1 Tbs. salt
- 1 Tbs. pepper
- 1½ c. bananas, peeled, chopped
- ½ c. shelled pistachios (optional)
- 1 can coconut milk (14 oz.)
- 1 garlic clove, crushed

Ingredients for rice:

- 2½ c. long-grain white rice
- 6 c. water
- 1 tsp. salt
- 1 onion, peeled
- 1 branch parsley
- 1 branch thyme
- 1 carrot, peeled

Directions for chicken:

1. In large saucepan, heat oil and butter.
2. Stir in chicken and Colombo powder.
3. Cook over medium heat until brown.
4. Add chicken broth, onions, ½ can of coconut milk, garlic, lime juice, thyme, chives, parsley, Colombo powder, salt, and pepper.
5. Simmer over medium heat, 45 minutes; turn off heat.
6. Stir in remaining coconut milk, bananas, and pistachios.

Directions for rice:

1. Soak rice in cold water for 15 minutes and drain.
2. In large saucepan, bring water to boil.
3. Add salt, onion, parsley, thyme, and carrot.
4. Add rice.
5. Simmer over medium heat, 20 minutes.
6. Remove onion, parsley, thyme, and carrot.
7. In colander, drain rice.
8. Rinse under cold running water.
9. Drain again and turn into saucepan.
10. Simmer over low heat, 5 minutes until rice grains are completely dry.

Yields: 6 servings.

Note for Colombo Powder: Colombo powder is a French, West Indies curry powder. It is a mixture of cumin, coriander, brown mustard, Malabar black pepper, cloves, fenugreek, and turmeric. Check gourmet shops and ethnic sections of your market, or order online. You may also experiment with making your own combination using spices listed above.

Chicken in Coconut Cream

This is a simple make-ahead meal. Serve with your favorite rice, salad, and dinner rolls.

Ingredients:

1 bundle Chinese cabbage or any green leafy vegetable
1 yam, peeled
2 potatoes or sweet potatoes, peeled
2 c. thick coconut cream
1 chicken, cut into pieces
2 onions, chopped
2 tomatoes, chopped
2 garlic cloves, chopped

Directions:

1. Preheat oven to 350 degrees F.
2. Wash and chop cabbage or other leaves.
3. Cut root vegetables into pieces.
4. Pour 1 cup of coconut cream into baking dish.
5. Arrange all ingredients in dish.
6. Top with remaining cream.
7. Bake for 2 hours.

Chicken Poached in Coconut Curry

Our family loves curry, and this is a simple one that tastes great. This is delicious served with steamed greens, a salad, and rice.

Ingredients:

3 Tbs. red curry paste
2 c. sweet potato, diced
2 c. chicken stock
1½ c. unsweetened coconut milk

¼ c. cilantro, chopped
3 chicken breasts, cut into pieces

Directions:

1. Place frying pan over medium-high heat, add curry paste.
2. Cook 1 to 2 minutes, until fragrant.
3. Add chicken and sweet potato, cook 2 minutes.
4. Add stock and coconut milk; reduce heat to low.
5. Do not boil.
6. Simmer 12 minutes, until chicken and sweet potato are cooked.
7. Sprinkle cilantro over chicken.

Fish in Coconut Cream

Mash or sieve this mixture for infants under 6 months of age. This is excellent baby food.

Ingredients:

2 Tbs. tinned or cooked fresh fish
4 Tbs. taro, cooked, mashed
4 Tbs. taro leaves, cooked, mashed
4 Tbs. thick coconut cream

Directions:

1. In saucepan, mix taro leaves with coconut cream, fish and mashed taro.
2. Cook for 5 minutes.
3. Cool and serve.

Yields: 1 serving.

Coconut Curry Thai Chicken

Coconut milk and red curry paste add a delightful spicy flavor to this wonderful dish. You may substitute shrimp or tofu for the chicken, and it is equally delicious.

Ingredients for chicken:

2 Tbs. curry powder
2 Tbs. peanut butter
1 c. roasted red peppers, thinly sliced
1 c. water chestnuts, thinly sliced
1 c. scallions, white and green parts, thinly sliced
8 oz. whole-wheat linguine
1 lb. boneless, skinless chicken breast, cut into small pieces
1 sweet onion, chopped
fresh mint and cilantro leaves, chopped, for garnish
lime wedges, for serving
kosher salt and freshly ground pepper
olive oil cooking spray

Ingredients for sauce:

1 Tbs. curry paste
2 Tbs. curry powder
2 tsp. fresh ginger, minced
1 can light coconut milk (13.5 oz.)
3 garlic cloves, minced
dash cayenne pepper
kosher salt and freshly ground pepper

Directions for chicken:

1. In large pot, add water and cook linguine according to directions on package.
2. Drain and rinse under cold water.
3. Spray lightly with oil; set aside.

4. Coat large nonstick skillet with cooking spray.
5. Add onion.
6. Cook 5 to 10 minutes, until translucent and beginning to brown.
7. Add chicken.
8. Season with salt and pepper; stir in curry powder.
9. Sauté chicken until light brown.
10. Add peanut butter; melt to coat chicken.
11. Add red peppers, water chestnuts, and scallions.
12. Season with salt and pepper.
13. Sauté for a few more minutes.
14. Add linguine to mixture.
15. Pour coconut sauce over chicken and linguine, stir gently.
16. Cook 3 to 5 minutes, until warmed through and thickened.
17. Season with salt and pepper.
18. Garnish with mint and cilantro.
19. Serve with lime wedges.

Directions for sauce:

1. In medium bowl, combine all sauce ingredients.
2. Whisk until thoroughly combined.
3. Set aside.

Did You Know? . . . .

Did you know that there are three types of coconuts that are used in making rings and earrings? They are known as palm nuts. The Pati, Dende, and Piasava palm nuts are also known as vegetable ivory. When tumbled and polished, they divulge the beautiful colors that lay hidden beneath their raw outer surface.

Coconut Shrimp with Fruity Mayonnaise

This is yet another version of the popular coconut shrimp to try.

Ingredients for mayonnaise:

1 egg yolk
2 Tbs. lemon juice
¾ c. olive oil
¼ c. mango chutney
1 tsp. mild or hot chili sauce

Ingredients for shrimp:

1 c. fresh bread crumbs
1 c. coconut, unsweetened
1½ lb. uncooked large shrimp, peeled, deveined
1 egg, beaten
canola oil, for deep-frying

Directions for mayonnaise:

1. In blender, combine yolk, lemon juice, salt, and pepper.
2. Blend until smooth.
3. With blender running, gradually add olive oil, mixing thoroughly.
4. Add chutney and chili sauce.
5. Mix well.

Directions for shrimp:

1. In bowl, combine bread crumbs and coconut.
2. Dip shrimp into egg, then into bread crumb mixture, pressing firmly.
3. Chill 15 minutes.

4. In large saucepan, heat oil over medium heat until hot.
5. Add shrimp; cook 3 to 5 minutes until tender and golden.
6. Drain on paper towels and serve with mayonnaise.

Coconut Chicken with Fresh Fruit

Try this unique chicken dish with your friends. You will enjoy this Caribbean touch.

Ingredients:

⅓ c. coconut, flaked
½ tsp. salt
2 Tbs. butter
8 lg. boneless chicken breasts, skinned
1 Tbs. fresh ginger, finely chopped, or ¾ tsp. ground
1 c. whipping cream
3 lg. bananas, quartered
1 lg. papaya, peeled, seeded, halved, sliced
lime wedges, for garnish

Directions:

1. Preheat oven to 300 degrees F.
2. Toast coconut for 10 minutes; set aside.
3. In large frying pan, brown chicken breasts in butter.
4. Sprinkle ginger, 3 tablespoons coconut, and salt over chicken.
5. Add cream, cover.
6. Cook over medium heat, 10 minutes or until done.
7. Remove chicken and arrange with fruit on a large platter.
8. Spoon sauce over chicken and sprinkle with remaining toasted coconut.
9. Garnish with limes to squeeze on top, if desired.

Coconut Shrimp with Peanut Sauce

Coconut and shrimp is a popular combination, and peanut sauce makes it even better.

Ingredients for peanut sauce:

¼ c. chicken broth
1½ c. creamy peanut butter
¼ c. cilantro, chopped
3 oz. unsweetened coconut milk
1 oz. lime juice
1 oz. soy sauce
1 Tbs. fish sauce or 2-3 anchovies, ground
1 Tbs. hot sauce
2 Tbs. garlic, chopped
1 Tbs. ginger, chopped

Ingredients for shrimp:

½ c. cornstarch
¼ tsp. salt
¼ tsp. fresh ground white pepper
¼ tsp. cayenne pepper
3 c. coconut, fresh or sweetened, shredded
4 egg whites
24 lg. shrimp, peeled, deveined, butterflied
canola or peanut oil, for frying
peanut sauce

Directions for peanut sauce:

1. In blender or food processor, purée chicken stock, coconut milk, lime juice, soy sauce, fish sauce, hot sauce, garlic, and ginger.
2. Add peanut butter and pulse to combine.
3. Fold in cilantro.
4. Keep refrigerated until ready to serve.
5. Bring sauce to room temperature to serve.

Directions for shrimp:

1. Pat shrimp dry with paper towel.
2. In small bowl combine cornstarch, salt, pepper, and cayenne pepper.
3. In separate bowl, whisk egg whites until foamy.
4. In another bowl, place coconut.
5. Coat shrimp with cornstarch and shake off excess.
6. Dip into egg white; press into coconut, cover well.
7. In large pan, heat oil to 350 degrees F.
8. Gently submerge shrimp, 6 at a time.
9. Fry 3 minutes or until golden brown.
10. Remove to rack to drain.
11. Serve with peanut sauce.

Prawns in Coconut Shells

This makes an attractive presentation for your next dinner party or special event.

Ingredients:

1 lb. prawns or shrimp (about 15)
1 c. water, salted
1 sm. green papaya (a.k.a. pawpaw)
2 mature coconuts
salt

Directions:

1. Preheat oven to 350 degrees F.
2. In large saucepan, boil prawns for 5 minutes in the salt water; drain, reserving water.
3. Shell and remove heads from prawns; chop flesh.
4. Grate coconuts and save shells.
5. Prepare coconut cream using prawn water.
6. Wash pawpaw thoroughly three times; remove seeds. Grate or chop into small pieces.
7. Put pawpaw in coconut shells with prawns and coconut cream. Cover with a banana leaf or grease-proof paper.
8. Bake 1 hour.

Chicken Meatballs with Coconut Curry

Try these delicious chicken meatballs with a coconut curry sauce. Enjoy!

Ingredients for meatballs:

1 chicken breast, cut into thin strips
½ chili bell pepper, chopped
2 Tbs. fresh chives, chopped
2 Tbs. fresh parsley, chopped
2 Tbs. olive oil

Ingredients for sauce:

1¾ oz. fresh coconut flesh, grated
½ lime, juice only
1 garlic clove, finely chopped
½ scotch bonnet chili, chopped
1 Tbs. fresh coriander, chopped
3½ oz. heavy cream
liquid from one fresh coconut
fresh coriander leaves, for garnish

Directions for meatballs:

1. Preheat oven to 350 degrees F.
2. In food processor or blender, place chicken, pepper, chives, and parsley; blend into a mince mixture.
3. Roll equal portions of the mince mixture into golf ball-size meatballs.
4. In medium skillet, heat oil; add meatballs; brown for 2 minutes. Transfer meatballs to a baking sheet.
5. Bake 10 minutes.

Directions for sauce:

1. In medium saucepan, over medium heat, add grated coconut, coconut liquid, lime juice, garlic, chili, coriander, and cream.
2. Cook until reduced to a thick sauce, then pour into a warm bowl.
3. Place meatballs in the bowl of sauce.
4. Garnish with a few coriander leaves.

Meatballs in Spicy Coconut Sauce

Our family loves meatballs, and this makes great-tasting meatballs with a unique sauce.

Ingredients for meatballs:

1 lb. ground round or ground pork
⅓ c. green onions, chopped
¼ c. water chestnuts, chopped
2 Tbs. cornstarch
1 Tbs. flour
1 Tbs. fresh ginger, peeled, minced
1 Tbs. low-sodium soy sauce
1 Tbs. dark sesame oil
1 tsp. hot red chili pepper, minced
¼ tsp. salt
1 Tbs. olive oil
¼ c. fresh basil, chopped
1 Tbs. lemon rind, grated

Ingredients for sauce:

½ c. coconut milk
½ c. soy milk
2 Tbs. fresh ginger, peeled, minced
2 tsp. hot red chili pepper, minced
1 Tbs. green onions, chopped
2 Tbs. fish sauce

Directions for meatballs:

1. In large bowl, combine ground pork and next 9 ingredients.
2. Shape mixture into 8 meatballs.
3. In a large, nonstick skillet, over medium-high heat, add 1 tablespoon olive oil.
4. Add meatballs.

5. Cook 7 minutes, browning on all sides.
6. Drain well.
7. Note: You can make meatballs a day or two ahead.
8. Store covered in refrigerator.
9. Heat meatballs in sauce before serving.

Directions for sauce:

1. In large saucepan, combine sauce ingredients mixing well.
2. Over medium-high heat, bring to a boil.
3. Add meatballs; cover.
4. Reduce heat, simmer 8 minutes.
5. Garnish with basil and rind.
6. Note: If making this sauce ahead of time, reheat and add meatballs. Heat thoroughly before serving.

Yields: 8 servings.

Did You Know? . . . .

Did you know that coconut jewelry has been made for generations by natives of the rainforests and coasts of South America?

Did you know that the Piasava coconut is the largest coconut that is used to make jewelry? It is used to make large hoop earring like Gypsy's wear.

Did you know that after wearing the jewelry for several weeks, the color will slowly begin to change to a darker brown?

Coconut Delights Cookbook

A Collection of Coconut Recipes
Cookbook Delights Series – Book 4

Pies

Table of Contents

Page

A Basic Recipe for Pie Crust

This is a very good recipe for a delicious, flaky crust.

Ingredients for single crust:

1½ c. sifted all-purpose flour
½ tsp. salt
½ c. shortening
4-5 Tbs. ice water

Ingredients for double crust:

2 c. sifted all-purpose flour
1 tsp. salt
⅔ c. shortening
5-7 Tbs. ice water

Directions for single crust:

1. In large bowl stir together flour and salt.
2. Cut in shortening with pastry blender or mix with fingertips until pieces are size of coarse crumbs.
3. Sprinkle 2 tablespoons ice water over flour mixture, tossing with fork.
4. Add just enough remaining water 1 tablespoon at a time to moisten dough, tossing so dough holds together.
5. Roll pastry into 11-inch circle, and wrap in plastic wrap; refrigerate for 1 hour.
6. Preheat oven to 425 degrees F.
7. Remove plastic wrap from pastry, and fit pastry into a 9-inch pie plate.
8. Fold edge under and then crimp between thumb and forefinger to make fluted crust.
9. For filled pie with an instant or cooked filling (cream-filled, custard-filled, etc.), prick crust all over with fork then bake 15 to 20 minutes until done.
10. If preparing pie with uncooked filling (such as pumpkin), do not prick crust; pour filling into unbaked pastry shell, and then bake as directed.

Directions for double crust:

1. Turn desired filling into pastry-lined pie plate; trim overhanging edge of pastry ½ inch from rim of plate.
2. Cut slits with knife in top crust for steam vents.
3. Place over filling; trim overhanging edge of pastry 1 inch from rim of plate.
4. Fold and roll top edge under lower edge, pressing on rim to seal; flute.
5. Cover fluted edge with 2- to 3-inch-wide strip of aluminum foil to prevent excessive browning.
6. Remove foil during last 15 minutes of baking.

Yields: 1 pie crust (9-inch single or double).

A Basic Cookie or Graham Cracker Crust

This is a great crust for use with cream pies or for an unbaked pie. Use your favorite flavor of cookie to complement your filling, or use graham crackers.

Ingredients:

2 c. cookie or graham cracker crumbs, finely crushed
⅓ c. sugar
½ c. butter, melted

Directions:

1. Combine crumbs, sugar, and butter.
2. Press mixture firmly against bottom and up sides of 9-inch pie plate.
3. Baking is not necessary, but if preferred crust may be baked at 400 degrees F. for 10 minutes.

Yields: 1 pie crust (9-inch).

Banana Coconut Pie

This is an easy-to-make, rich dessert.

Ingredients for graham cracker crust:

1¼ c. graham cracker crumbs (8 graham crackers)
¼ c. sugar
⅓ c. butter, melted

Ingredients for pie filling:

3 Tbs. cornstarch
1½ c. cold water
1 can sweetened condensed milk (14 oz.)
4 egg yolks, beaten
2 Tbs. butter
1 tsp. vanilla extract
1 c. coconut, flaked
1½ med. banana, sliced
1 graham cracker pie crust

Directions for crust:

1. Preheat oven to 375 degrees.
2. In small bowl, mix all ingredients until well blended.
3. Press firmly onto bottom and up side of 9-inch pie plate.
4. Bake 8 to 10 minutes, until lightly browned.
5. Cool completely.

Directions for pie:

1. In saucepan, whisk together cornstarch, water, milk, and egg yolks.
2. Over medium heat, cook and stir until thick and bubbly.

3. Remove from heat.
4. Whisk in butter and vanilla.
5. Cool slightly.
6. Fold in coconut.
7. Slice banana and arrange in bottom of crust.
8. Pour filling over bananas.
9. Cover with plastic wrap.
10. Chill 4 hours or until set.
11. Garnish with toasted coconut.

Buttermilk Coconut Pie

This is a nice change for coconut lovers.

Ingredients:

⅓ c. butter, melted
1 c. sugar
2 Tbs. flour
5 eggs, beaten lightly
1 c. buttermilk
2 tsp. vanilla extract
1½ c. coconut, grated
dash of salt

Directions:

1. Preheat oven to 350 degrees F.
2. In medium bowl, add butter.
3. Stir in sugar, flour, and salt.
4. Add beaten eggs, vanilla, and buttermilk.
5. Stir until well blended.
6. Stir in coconut.
7. Pour into unbaked pie shell.
8. Bake 45 minutes, until set and lightly browned.
9. Serve warm or chilled.
10. Garnish with whipped cream and coconut.

Blueberry Coconut Pie

Blueberries complement the coconut in this elegant creamy pie.

Ingredients:

1 9-inch pastry shell, unbaked
1½ c. coconut, flaked
¼ c. walnuts, chopped
¼ c. light corn syrup
1 egg, well beaten
1 Tbs. flour
¼ tsp. salt
¼ c. sugar
1 pkg. blueberries, frozen, unsweetened (10 oz.)
⅔ c. sugar
1 c. heavy cream, whipped (½ pt.)

Directions:

1. Preheat oven to 425 degrees F.
2. Make pastry shell from your own recipe or use a mix.
3. Bake as directed for 5 minutes.
4. Reduce heat to 375 degrees.
5. In medium bowl, combine egg, coconut, nuts, syrup, flour, salt, and ¼ cup sugar.
6. Spread in bottom of partially baked pastry shell.
7. Return pie to oven.
8. Bake 15 minutes.
9. Cool thoroughly.
10. In small bowl, crush frozen blueberries.
11. Combine with ⅔ cup sugar.
12. Fold into whipped cream.
13. Pour berry mixture over cooled coconut mixture and freeze.
14. Note: Use 2 cups fresh blueberries, when in season.
15. Chill berry mixture before folding into cream.

Yields: 6 servings.

Chocolate Coconut Lime Pie

The flavors in this pie blend to make a delicious dessert for family or friends.

Ingredients:

1 Tbs. plain gelatin
1 cup sugar, divided
¼ tsp. salt
5 eggs, separated
½ c. lime juice, strained
¼ c. water
1 tsp. lime peel, grated
½ c. coconut, shredded
1 c. heavy cream, whipped
1 9-inch baked chocolate graham cracker crust
grated chocolate and lime slices, for garnish

Directions:

1. In small saucepan, mix gelatin, ½ cup of sugar, and salt.
2. In small bowl, beat egg yolks.
3. Add lime juice and water to egg yolks.
4. Stir egg yolk mixture into gelatin mixture.
5. Cook on low heat, stirring, just until mixture comes to a boil.
6. Remove from heat; blend in grated peel, and coconut.
7. Chill, blending often, until mixture is thick when dropped by spoon.
8. In small bowl, beat egg whites until soft peaks form.
9. Add remaining sugar and beat until stiff.
10. Fold into gelatin mixture.
11. Fold in whipped cream.
12. Pour into crust; refrigerate until firm.
13. Serve with a sprinkle of chocolate and garnish with lime slice.

Chocolate Coconut Macaroon Tarts

The flavors of these little cups of toasted coconut are delicious and easy to make. Enjoy!

Ingredients:

¾ c. sugar
½ c. egg whites (from 3 eggs)
2½ c. coconut, sweetened, flaked or desiccated
½ c. heavy cream
8 oz. semi-sweet chocolate, chopped
24 mini muffin cups or individual tart molds
few toasted almonds, chopped, for garnish

Directions for filling:

1. In small bowl, place chopped chocolate.
2. In small saucepan, heat cream until just boiling.
3. Pour over chocolate; let sit for 1 minute.
4. Whisk gently until glossy to melt chocolate completely.

Directions for tarts:

1. Preheat oven to 350 degrees F.
2. In medium bowl, mix sugar, egg whites, and coconut together.
3. Put a spoonful in each cup or mold.
4. Press into the molds to make little cups, with sides and a well for holding the chocolate filling.
5. Bake 30 minutes, until golden.
6. Cool completely in pans; remove gently.
7. Make the filling and pour over tarts.
8. Sprinkle a few pieces of chopped almonds in the center of each tart while warm.
9. Let set at room temperature, 1 hour before serving.
10. Note: Chocolate can be refrigerated up to 5 days.

11. Tart shells can be baked up to 2 days in advance and kept at room temperature in an airtight container.
12. Rewarm in microwave or bowl set over simmering water and pour over tarts.

Huckleberry Coconut Pie

This huckleberry pie keeps well in your freezer to pull out for surprise guests or for your invited company.

Ingredients:

- 1 egg, well beaten
- 1½ c. coconut, flaked
- ¾ c. walnuts, chopped
- ¼ c. light corn syrup
- 1 Tbs. flour
- ¼ c. sugar
- 1 pkg. huckleberries, frozen, unsweetened (10 oz.)
- ⅔ c. sugar
- 1 c. heavy cream, whipped
- 1 9-inch pastry shell, unbaked

Directions:

1. Preheat oven to 375 degrees F.
2. Make pastry shell from your own recipe or use pre-made pastry; bake as directed for 5 minutes.
3. In medium bowl, combine egg with coconut, nuts, syrup, flour, salt, and ¼ cup sugar.
4. Spread in bottom of pastry shell.
5. Return pie to oven, bake 15 minutes.
6. Cool thoroughly.
7. In small bowl, crush frozen huckleberries; combine with ⅔ cup sugar.
8. Fold into whipped cream.
9. Pour berry mixture over coconut mixture; freeze.

Coconut Chocolate Pie

This pie is a chocolate coconut lover's dream. It is very rich and delicious.

Ingredients:

- 1 unbaked pie crust, 9-inch
- ½ c. cornstarch
- 1 c. semi-sweet chocolate chips
- 1½ c. heavy cream
- ¼ c. sugar
- 1 c. milk
- 1 can coconut milk (14 oz.)
- 1 c. sugar
- 1 c. water
- toasted coconut and shaved chocolate, for garnish

Directions:

1. Preheat oven to 350 degrees F.
2. Bake crust 15 minutes, or until golden brown.
3. Set aside to cool.
4. In medium saucepan, whisk together milk, coconut milk, and 1 cup sugar.
5. In separate bowl, dissolve cornstarch in water.
6. Bring coconut mixture to a boil.
7. Reduce to simmer, slowly whisk in cornstarch.
8. Continue stirring over low heat until thickened, 3 minutes.
9. In glass bowl, microwave chocolate chips for 1 minute or until melted.
10. Divide coconut mixture evenly into two bowls.
11. Mix chocolate into one portion.
12. Spread on bottom of pie crust.
13. Pour remaining portion of coconut mixture on top of chocolate, spread smooth.
14. Refrigerate 1 hour.

15. In medium bowl, whip cream with ¼ cup sugar until stiff peaks form.
16. Layer cream on pie.
17. If desired, garnish with toasted coconut and chocolate shavings or chocolate curls.

Coconut Oatmeal and Pecan Pie

In this rich pie, the coconut and oatmeal complement each other in flavor and texture.

Ingredients:

- 3 eggs, lightly beaten
- 1 c. corn syrup
- ⅔ c. sugar
- ⅓ c. butter, melted
- 1 tsp. vanilla extract
- ¾ c. coconut, shredded
- ½ c. quick-cooking rolled oats
- ¾ c. pecans, chopped
- 12 shelled pecans, unbroken, for garnish
- 1 9-inch pie shell

Directions:

1. Preheat oven to 350 degrees F.
2. In large bowl, beat eggs and butter.
3. Add all ingredients except whole pecans.
4. Pour into pastry-line pie plate.
5. Cover edge of pie with aluminum foil.
6. Bake 25 minutes.
7. Remove foil.
8. Bake for 20 to 25 minutes more, or until inserted toothpick comes out clean.
9. Place whole pecans around edge of pie, for garnish.

Yields: 8 servings.

Coconut Cream Pie

My family loves coconut cream pie, and my mom used to make it on a regular basis for all of us to enjoy.

Ingredients for graham cracker crust:

1¼ c. graham cracker crumbs (8 graham crackers)
¼ c. sugar
⅓ c. butter, melted

Ingredients for pie:

⅔ c. sugar
¼ c. cornstarch
½ tsp. salt
3 c. milk
5 egg yolks, slightly beaten
2 Tbs. butter, softened
2 tsp. vanilla extract
1⅓ c. coconut
2 c. sweetened whipped cream, for garnish
1 graham cracker crust

Directions for crust:

1. Preheat oven to 375 degrees.
2. In small bowl, mix all ingredients until well blended.
3. Press firmly onto bottom and up side of 9-inch pie plate.
4. Bake 8 to 10 minutes, until lightly browned.
5. Cool completely.

Directions for pie:

1. In small bowl, add egg yolks; beat slightly.

2. In medium saucepan, mix sugar, cornstarch, and salt.
3. Gradually stir in milk; cook over medium heat, stirring constantly until mixture thickens and boils.
4. Boil and stir, 1 minute.
5. Stir at least half of hot mixture gradually into yolks.
6. Stir yolk mixture into hot mixture in saucepan.
7. Boil and stir, 1 minute.
8. Remove from heat.
9. Stir in butter, vanilla, and ¾ cup of coconut.
10. Pour into shell.
11. Cover with plastic wrap and refrigerate at least 2 hours, but no longer than 48 hours.
12. Remove plastic wrap; top pie with sweetened whipped cream and remaining coconut.
13. Refrigerate any remaining pie immediately.

Sherbet Pie in Coconut Crust

This is a simple and quick pie to make. The coconut pie crust adds a nice touch to the sherbet.

Ingredients:

3 Tbs. butter, softened
1½ c. coconut, flaked
4 c. lime, raspberry, orange or lemon sherbet

Directions:

1. Preheat oven to 325 degrees F.
2. Grease pie pan with 3 tablespoons butter.
3. Press coconut firmly and evenly against bottom and sides of pan.
4. Bake 15 to 20 minutes, until golden brown.
5. Cool.
6. Scoop sherbet into crust.
7. Serve immediately.

Coconut Cream Meringue Pie

Try this creamy and delicious coconut pie. The meringue makes a great topping.

Ingredients:

¾ c. sugar
¼ c. cornstarch
¼ tsp. salt
1 tsp. vanilla extract
2 Tbs. butter
2 c. milk
3 egg yolks
1 c. coconut, flaked
1 9-inch pie shell, baked

Ingredients for meringue:

4 egg whites
6 Tbs. sugar

Directions for pie:

1. In a heavy saucepan, combine sugar, cornstarch, salt, and milk.
2. Over medium-high heat, cook stirring constantly, until thick and bubbly.
3. Boil one minute; remove from heat.
4. In medium bowl, beat 3 egg yolks.
5. Gradually stir in ¼ of the hot mixture into yolks.
6. Pour yolks back into remaining hot mixture, stirring constantly.
7. Cook, stirring constantly for 30 seconds.
8. Remove from heat.
9. Add butter, vanilla, and coconut.
10. Pour into baked shell.
11. Cool in refrigerator.

Directions for meringue:

1. In medium bowl, beat egg whites until frothy.
2. Continue to beat while slowly adding 6 tablespoons of sugar.
3. Beat until whites form stiff peaks.
4. Spread meringue over pie and sprinkle with coconut.
5. Brown in hot oven in hot oven for 5 minutes, or until light brown.

Meringue

I love meringue, and it is great on many of your favorite pies.

Ingredients:

½ tsp. cream of tartar
½ c. sugar
5 lg. egg whites, room temperature

Directions:

1. Preheat oven to 350 degrees F.
2. In mixing bowl, combine egg whites and cream of tartar.
3. With electric mixer on medium speed, beat until soft peaks form.
4. Gradually add sugar, 1 tablespoon at a time.
5. Beat on high speed, 3 minutes, or until mixture forms stiff, glossy peaks.
6. Immediately spread meringue over pie, carefully sealing to the edge of the pastry.
7. Bake 12 to 15 minutes, until light golden brown.
8. Cool on wire rack.

Coconut Cream Pie with Coconut Milk

Toasted coconut and coconut milk add great flavor to this coconut cream pie. Try this delicious recipe full of coconut.

Ingredients for crust:

- 2 c. flour
- 1 Tbs. sugar
- 1 tsp. salt
- ¾ c. solid vegetable shortening
- 6-7 Tbs. ice cold water

Ingredients for pie:

- ¾ c. sugar
- 1½ c. unsweetened coconut milk
- 1½ c. whole milk
- ¼ c. cornstarch
- 6 egg yolks
- 1⅓ c. coconut, flaked, unsweetened
- 2 tsp. vanilla extract
- 2 Tbs. butter
- ½ c. toasted coconut
- 1 9-inch baked pie shell
- whipped cream, for garnish

Directions for crust:

1. Preheat oven to 350 degrees F.
2. In mixing bowl, combine flour, sugar, and salt.
3. Mix well.
4. Cut in shortening, and mix until mixture resembles coarse crumbs.
5. Add water; let sit 1 minute.
6. Using a fork or hands, press mixture together to form a soft ball.

7. Wrap in plastic wrap, refrigerate at least 30 minutes.
8. Remove dough from refrigerator and place on a lightly floured surface.
9. Roll dough into a circle, 12 inches in diameter and ¼-inch thick.
10. Gently fold the dough in half and then in half again.
11. Unfold into a 9 x 2-inch deep-dish pie pan.
12. Crimp edges.
13. Chill again for 30 minutes before baking.
14. Bake 25 to 30 minutes, until golden.
15. Remove from oven.
16. Cool completely.

Directions for pie:

1. In 1-quart saucepan, combine ¾ cup sugar, coconut milk, and 1 cup milk.
2. Scald mixture.
3. In small mixing bowl, whisk the remaining ½ cup milk and cornstarch.
4. In medium bowl, whisk egg yolks and salt.
5. Add ½ cup scalded milk mixture to yolks and whisk well.
6. Add yolk and cornstarch mixtures into milk mixture.
7. Over medium heat, whisk vigorously until thickened, about 2 minutes.
8. Remove from heat.
9. Add coconut, vanilla, and butter.
10. Whisk until well mixed together.
11. Pour the filling into prepared pie shell.
12. Cover with plastic wrap and place in refrigerator.
13. Chill pie completely, 2 hours.
14. Cut into wedges.
15. Top with toasted coconut and a dollop of whipped cream, if desired.

Coconut Lemon Pie

This makes a refreshing combination of lemon and coconut.

Ingredients:

- 2 c. water
- ½ c. lemon juice
- ¾ tsp. salt
- 1 c. sugar
- 1 c. water
- 1 c. flour
- ½ c. sugar
- 1 Tbs. lemon rind, grated
- 2 Tbs. butter
- 10 oz. coconut, shredded
- 5 egg yolks
- ¼ c. coconut, toasted, for garnish
- 1 baked 9-inch pie shell

Directions:

1. In top of double boiler, combine 2 cups water, lemon juice, salt, and 1 cup sugar.
2. Bring to boil over direct heat.
3. Place over boiling water.
4. Add 1 cup water gradually to flour, mixing to a smooth paste.
5. Add this gradually to hot mixture, cook 5 minutes, stirring constantly.
6. Beat egg yolks with ½ cup sugar.
7. Add slowly to lemon mixture.
8. Cook 2 minutes, stirring constantly.
9. Add lemon rind and butter.
10. Cool 10 minutes.
11. Pour into pie crust.
12. Sprinkle with toasted coconut for extra flavor.

Rhubarb Coconut Pie

The coconut topping adds a nice change to traditional rhubarb pie.

Ingredients for crust:

1 c. flour
½ c. butter
1 tsp. baking powder
1 tsp. milk
1 egg

Ingredients for filling:

2½ c. rhubarb, cut into 1-inch pieces
2 Tbs. water
⅔ c. sugar, plus 2 Tbs.
1 Tbs. cornstarch
2 Tbs. cold water

Ingredients for topping:

1 c. sugar
½ c. butter
1 tsp. vanilla extract
2 c. coconut, grated
1 egg

Directions for crust:

1. Lightly grease a 9-inch pie pan.
2. In medium bowl, sift together flour and baking powder.
3. Cut in butter with a knife until texture of coarse crumbs.
4. In small bowl, beat egg and milk together.
5. Stir into flour mixture until well blended.
6. Pat into bottom of prepared pan.
7. Pour cooled rhubarb sauce over crust.

Directions for filling:

1. In small saucepan, combine rhubarb and sugar plus 2 tablespoons.
2. Cook over medium heat until rhubarb is runny, stirring occasionally to prevent scorching.
3. In small bowl, mix together cornstarch and cold water.
4. Add to rhubarb, cook until sauce clears and thickens; set aside to cool.
5. Preheat oven to 350 degrees F.

Directions for topping:

1. In small bowl, cream together sugar and butter.
2. Beat in vanilla and egg; stir in coconut.
3. Spread over filling.
4. Bake 30 minutes; cool before cutting.

Quick Blend Coconut Custard Pie

This makes a quick and easy blender pie when you want a quick coconut dessert.

Ingredients:

⅓ c. butter
½ c. flour
¾ c. sugar
1½ tsp. vanilla extract
2½ c. milk
4 eggs
1½ c. coconut

Directions:

1. Preheat oven to 350 degrees F.
2. Combine all ingredients in blender; blend 1 minute.
3. Pour into well greased and floured 10-inch pie pan.
4. Bake 40 to 45 minutes, until custard sets.

Coconut Delights Cookbook
A Collection of Coconut Recipes
Cookbook Delights Series – Book 4

Preserving

Table of Contents

Page

A Basic Guide for Canning, Dehydrating, and Freezing

1. Place empty jars in hot, soapy water. Wash well inside and out with brush or soft cloth.
2. Run your finger around rim of each jar, discarding any that are chipped or cracked.
3. Rinse in clean, clear, very hot water, being careful to use tongs to avoid burning skin or fingers.
4. Place upside down on towel or fabric to drain well.
5. Place lids in boiling water bath for 2 minutes to sterilize and keep hot until ready to place on jar rims.
6. Immediately prior to filling jars with hot food, immerse in hot bath for 1 minute to heat jars. Heating jars avoids breakage.
7. If filling with room-temperature food, you need not immerse immediately prior to filling.
8. Fill jars with food to within ½ inch of neck of jars.
9. When ladling liquid over food, fill jars to 1 inch from top rim in each jar. This leaves air allowance for sealing purposes.
10. Wipe rims of jars with damp, clean cloth to remove any particles of food and again check for chips or cracks.
11. Using tongs, place lids from hot bath directly onto jars.
12. Place rings over lids, and using cloth, gloves, or holders, tighten down firmly while hanging onto jars.
13. Do not tighten down too hard as air may become trapped in jars and prevent them from sealing.
14. For fruits, tomatoes, and pickled vegetables, place each jar into water bath canning kettle so water covers jars by at least 1 inch.
15. For vegetables, process them in a pressure canner according to manufacturer's directions.
16. Follow time recommended for food being canned.
17. Do not mix jars of food in same canning kettle as times may vary for each kind of food.

18. At end of time recommended for canning, gently lift each jar out of bath with tongs, and place on protected surface.
19. Turn lids gently to be sure they are firmly tight.
20. Place filled, ringed jars on cloth to cool gradually.
21. Do not disturb rings, lids, or jars until sealed.
22. Lids will show slight indentation when sealed.
23. When cool, wipe jars with damp cloth then label and date each jar.
24. Leave overnight until thoroughly cooled.
25. Jars may then be stored upright on shelves.

Dehydrating

1. Always begin with fresh, good quality food that is clean and inspected for damage.
2. Pretreatment is not necessary, but food that is blanched will keep its color and flavor better. Use the same blanching times as you would for freezing. Fruit, especially, responds to pretreatment.
3. Doing some research on pretreatments may help you decide what procedure you would like to use.
4. You can marinate, salt, sweeten, or spice foods before you dehydrate them.
5. Jerky is meat that has been marinated and/or flavored by rubbing spices into it; avoid oil or grease of any kind as it will turn rancid as the food dries.
6. Vegetables and fruit can be treated the same way.
7. Slice or dice food thin and uniform so that it will dehydrate evenly. Uneven thicknesses may cause food to spoil because it did not dry as thoroughly as other parts.
8. Space food on dehydrator tray so that air can move around each piece.
9. Try not to let any piece touch another.
10. Fill your trays with all the same type of food as different foods take different amounts of time to dry.

11. You can, of course, dry different types of food at the same time, but you will have to remember to watch and remove the food that dehydrates more quickly. You can mix different foods in the same dehydrator batch, but do not mix strong vegetables like onions and garlic as other foods will absorb their taste while they are dehydrating.
12. The smaller the pieces, the faster a food will dehydrate. Thin leaves of spinach, celery, etc., will dry fastest. Remove them from the stalks before drying them or they will be overdone, losing flavor and quality. In very warm areas, they might even scorch. If they do, they will taste just like burned food when you rehydrate them.
13. Dense food like carrots will feel very hard when they are ready. Others will be crispy. Usually, a food that is high in fructose (sugar) will be leathery when it is finished dehydrating.
14. Remember that food smells when it is in the process of drying, so outdoors or in the garage is an excellent place to dry a big batch of those onions!
15. Always test each batch to make sure it is "done."
16. You can pasteurize finished food by putting it in a slow oven (150 degrees F.) for a few minutes.
17. Let the food cool before storing.
18. Store in airtight containers to guard against moisture. Jars saved from other food work well as long as they have lids that will keep moisture out.
19. Zip-closure food storage bags work well.
20. Jars of dehydrated carrots, celery, beets, etc., may look cheerful on your countertop, but the colors and flavors will fade. Dehydrated food keeps its color and flavor best if stored in a dark, cool place.
21. Dehydrating food takes time, so do not rush it. When you are all done, you will have a dried food stash to be proud of!

Freezing

1. Wash all containers and lids in hot, soapy water using soft cloth.
2. Rinse well in clear, clean, hot water.
3. Cool and drain well.
4. Place food into container to within 1 inch of rim. This allows for expansion of food during freezing.
5. Wipe rim of container with clean damp cloth, checking for chips or breaks.
6. Be certain cover fits the container snugly to avoid leaks. Burp air from container.
7. If food is hot when placing in container, cool prior to placing in freezer.
8. Label and date each container.
9. Store upright in freezer until frozen solid.

Frozen Coconut

It is always great to have some frozen coconut meat in your freezer to pull out for treats.

Ingredients:

1 qt. coconut, cut into pieces
3 lb. dry ice

Directions:

1. Cover coconut with paper towel; refrigerate 4 hours.
2. Break dry ice into small pieces; toss with coconut pieces in a large bowl.
3. Place into a container and cover with a towel.
4. Place in a cooler for 25 to 30 minutes.
5. Remove pieces and put into sealable bags and store in freezer.

Coconut Olive Soap

Coconut oil and olive oil combine to make a great soap.

Ingredients:

8 oz. olive oil
8 oz. coconut oil
8 oz. rendered tallow
3.49 oz. sodium hydroxide (pure lye)
9 oz. water

Directions:

1. Wearing safety goggles and neoprene gloves, combine solid lye and liquid, stir well.
2. Set aside.
3. Allow to cool, 100 – 125 degrees F.
4. Combine oils and heat gently.
5. Once the fats and oils are melted allow the temperature to drop to 100 – 125 degrees F.
6. Combine lye solution and melted oils.
7. Note: Be careful not to splash while combining the mixtures.
8. Stir until mixture traces. Tracing looks like a slightly thickened custard, not instant pudding but a cooked custard. It will support a drop, or your stir marks for several seconds.
9. If tracing takes more than 15 minutes, which it often does, stir for the first 15 minutes, then stir for 5 minutes at 15 minute intervals.
10. Pour raw soap into your prepared molds.
11. After a few days the soap can be turned out of the mold.
12. If the soap is very soft, allow it to cure for a few days to firm the outside.
13. Cut soap into bars; set the bars out to cure and dry.
14. This will allow the bar to firm and finish.

15. Place the bars on something that will allow them to breathe.
16. Note: If you do not want to use tallow and lye, you can substitute shavings from any white unscented soap.

Coconut Fruit Leather

My children always love homemade fruit leather, and this is a nice variation.

Ingredients:

coconut, grated
any ripe fruits, skin and all, except pit or seeds
cinnamon
lemon

Directions:

1. In food processor, purée any ripe fruit.
2. Turn a 7 x 11-inch glass dish upside down.
3. Cover with plastic wrap.
4. Spread ¼ cup purée with palm of hand into a circle, very thin with edges a little thicker.
5. Make a hole in center the size of a thumb print.
6. Microwave on 50% power, 8 to 12 minutes, or until slightly sticky to touch or follow directions to your food dehydrator.
7. Remove plastic wrap and leather.
8. Cool.
9. Repeat with remaining mixture.
10. Finish drying overnight.
11. Roll up.
12. Note: This fruit leather will keep for years.

Coconut Oil Soap

Mixing the ingredients in the blender is a unique way to make soap. This recipe will only make one pound batches.

Ingredients:

- 4 oz. coconut oil
- 4 oz. palm oil
- 8 oz. lard
- 2.3 oz. lye
- 7 oz. water
- 1 tsp. coconut essential or fragrance oil

Directions:

1. Put goggles on.
2. You will need a towel to cover blender.
3. In a bowl, dissolve lye in cold water, wait until it cools and turns clear.
4. In blender container, carefully pour oil; THEN add the lye/water solution. ** BE CAREFUL NOT TO SPLASH OR SPILL THE LYE ON YOURSELF OR OTHERS! **
5. Lock blender in position, secure cover.
6. Place towel over top of blender for safety.
7. Process at lowest possible speed.
8. Stop blender often and check soap, watching for what is called a thin-trace stage. This is when the soap mixture just begins to thicken.
9. Each time you stop the blender, wait a few seconds before removing the cover. Sometimes the soap "burps" when it stops as trapped air comes to the top.
10. At the thin trace stage, stop the blender and stir the soap to check for tracing and to allow bubbles to escape.
11. Add any essential oils, colorants or fragrances as well as any other ingredients such as oatmeal or herbs.
12. Blend for a few seconds; pour soap into individual molds.
13. Cover molds with a blanket for insulation.

14. Let soap set for a day or two; take out of molds.
15. Cut into pieces.
16. Let age for at least three weeks.

Toasted Coconut Chips

These coconut chips are actually very good as a snack or an appetizer. Enjoy!

Ingredients:

1 mature coconut

Directions:

1. Cut coconut in half; place coconut in hot sun, solar dryer or copra dryer.
2. When partially dry, remove coconut flesh from shell; remove thin brown peel.
3. Slice very thin and spread in shallow baking pan.
4. Place in hot sun or in solar dryer.
5. Dry 1 to 2 days or until no moisture is left; stirring several times.
6. Store in airtight container.

Roasted Coconut

This coconut will keep for a couple of months in a sealed jar in the pantry.

Ingredients:

coconut, shredded

Directions:

1. Spread shredded coconut in a dry pan over moderate heat.
2. Stir frequently until coconut turns a rich and even golden brown color and very fragrant.
3. Remove from pan.
4. Set aside to cool before using.

Coconut Oil

This oil is a soothing relief for head and chest colds. Rub on your chest and back before going to bed.

Ingredients:

1 coconut
rosemary

Directions:

1. Grate coconut.
2. Add 2 cups of warm water.
3. Squeeze milk from coconut.
4. Put in pot, boil until it becomes oil.
5. Put oil in a bottle and add rosemary to it.

Dried Coconut

Dried coconut is great for adding to recipes or for snacks. It is a convenient and healthy alternative to pack in lunches or take along as a treat.

Ingredients:

coconuts, washed, hulled

Directions:

1. Arrange on dehydrator tray, without touching.
2. Dry at 120 degrees F. for 12 to 24 hours.
3. Rotate the trays after about 10 hours.
4. When the pieces break easily they are dried.
5. For storage, vacuum seal and store in the freezer, or place in jars and screw lids on tightly.

Salads

Table of Contents

Apple Squash and Coconut Salad

This unusual flavor combination makes a salad with pizzazz.

Ingredients:

1¾ oz. cashews, toasted
1 Tbs. coriander, freshly chopped
¼ c. coconut, shredded, toasted
1 tsp. ginger, grated
1 c. coconut milk
½ c. lemon juice
2 tsp. soy sauce
1 butternut squash, peeled, cut into cubes
red apples, diced

Directions:

1. Steam, boil or microwave squash cubes until tender.
2. In serving bowl, combine squash, apples, cashews, coriander, and coconut.
3. In separate bowl, place ginger, coconut milk, lemon juice, and soy sauce; mix well.
4. Pour dressing over salad.
5. Toss well to combine.

Chunky Fruit and Coconut Salad

Try this delicious fruit salad.

Ingredients for salad:

1 c. pecans, chopped
4 med. oranges, peeled, seeded, sectioned
2 med. apples, unpeeled, chopped
2 med. bananas, peeled, sliced
1 c. seedless green grapes, halved

½ c. coconut, flaked
1 can pineapple chunks, drained (8 oz.)

Ingredients for sweetened whipped cream:

½ c. whipping cream
1½ Tbs. sugar
⅛ tsp. almond extract

Directions for salad:

1. In large bowl, combine salad ingredients.
2. Toss gently.
3. Chill salad before serving.

Directions for sweetened whipped cream:

1. In small bowl, beat whipping cream until foamy.
2. Gradually add sugar, beating until soft peaks form.
3. Stir in almond extract.
4. Fold whipped cream mixture into fruit.

Five Cup Ambrosia Salad

This salad is delicious and very easy and quick to make. My mom used to make it, and it always disappeared quickly.

Ingredients:

1 c. coconut, shredded
1 c. pineapple, chopped
1 c. canned mandarin orange slices
1 c. sour cream
1 c. miniature marshmallows

Directions:

1. Mix together and serve.

Coconut Ambrosia Salad

This salad is very popular with children as well as adults. It is also very easy to make.

Ingredients:

¾ c. coconut, flaked, toasted
¾ c. pecans, chopped, toasted
1 can pineapple chunks, drained (20 oz.)
1 can mandarin oranges drained (15 oz.)
1 jar maraschino cherries, drained, halved (6 oz.)
2 med. bananas, sliced
1 c. miniature marshmallows
1 c. sour cream (8 oz.)
2 Tbs. fresh lemon juice
1 Tbs. sugar

Directions:

1. In small bowl, put sour cream.
2. Stir in lemon juice and sugar; set aside.
3. In large bowl, add remaining ingredients.
4. Fold in sour cream mixture.
5. Cover and chill until serving time.
6. Note: If you make this salad early in the day, add bananas just before serving.
7. You can also substitute dark sweet pitted cherries for maraschino cherries.

Coconut Carrot Salad

Coconut goes very well with carrots. Try this unique salad with your favorite dinner for a change of pace..

Ingredients:

1 c. apple, finely chopped

1 c. coconut, flaked
6 tsp. honey
1 tsp. lemon juice
1 med. carrot
8 Tbs. sour cream

Directions:

1. In bowl, combine all ingredients; mix well.
2. Cover.
3. Refrigerate until serving.

Coconut Celery and Apple Salad

This makes a crisp, crunchy salad with the flavor of orange and lemon.

Ingredients:

¾ c. coconut, flaked
¼ c. olive oil
⅓ c. orange juice
1 Tbs. lemon juice
2-3 apples, unpeeled, diced
1 c. celery, diced
salt, to taste
paprika
lettuce leaves

Directions:

1. In large bowl, combine apples, celery, and coconut.
2. Sprinkle lemon juice over fruit.
3. Combine oil, orange juice, salt, and paprika.
4. Add to fruit; chill.
5. Line a salad bowl with lettuce leaves and pile chilled salad in center.

Coconut Waldorf Salad

My mom used to make traditional waldorf salad for my dad. This is a nice variation.

Ingredients:

1 c. walnuts, chopped
¾ c. coconut mayonnaise (see page 184)
3 apples, diced
3 ribs celery, chopped

Directions:

1. In small bowl, combine all ingredients.
2. Chill and serve.

Coconut Fruit Salad

This is an easy salad to make and always seems to disappear quickly.

Ingredients:

1 can mandarin oranges, drained (11 oz.)
2 c. seedless green grapes
1½ c. miniature marshmallows
1½ c. coconut, flaked
1 c. sour cream

Directions:

1. In large bowl, combine first 4 ingredients.
2. Fold in sour cream.
3. Chill several hours or overnight.
4. Serve on lettuce cups.

Yields: 8 servings.

Thai Slaw and Dressing

Thai seasoning makes this salad very flavorful.

Ingredients for dressing:

1 Tbs. fish sauce
¼ c. lime juice
2 tsp. sesame oil
1 tsp. red chili paste
½ c. coconut milk
½ c. peanut butter
red pepper flakes, sprinkle to taste

Ingredients for slaw:

2 c. red cabbage, thinly sliced
2 carrots, shaved or peeled
½ c. daikon, cut into matchsticks
½ med. red onion, thinly sliced
3 Tbs. cilantro, chopped

Directions for dressing:

1. In small pot, over medium-low heat, combine fish sauce, lime juice, red chili paste, milk, and peanut butter.
2. Cook, stirring often, 5 minutes.
3. Taste; add red pepper flakes if more spice is needed.

Directions for slaw:

1. In medium bowl, combine cabbage, carrots, daikon, red onion, and cilantro.
2. Mix in half of dressing.
3. Serve the remainder of the dressing with your favorite fish, shrimp, pork, or chicken recipe.

Coconut Mango Rice Salad

Cilantro and mint add a nice touch to this coconut-flavored rice salad.

Ingredients:

1½ c. English cucumber, chopped
1½ c. long-grain white rice
½ c. fresh parsley leaves, loosely packed
1½ tsp. sea salt
¼ c. red onions, chopped
1 can coconut milk (14 oz.)
1 Tbs. olive oil
2 med. ripe peaches, chopped
1 med. red pepper, chopped
1 med. mango, peeled, pitted, chopped
1 garlic clove, minced
1 Tbs. sugar
¼ c. red wine vinegar
1 c. olive oil
¼ tsp. sea salt
¼ tsp. fresh ground pepper
5 green onions, sliced
3 garlic cloves, minced
½ c. cilantro leaves, loosely packed
½ c. mint leaves, loosely packed

Directions:

1. In food processor, combine ¼ of sliced green onions, garlic, cilantro, mint, and parsley.
2. Pulse to finely chop; scrape into glass bowl.
3. Add coconut milk, mixing well.
4. Measure and add enough water to make 4 cups.
5. Whisk well and set aside.
6. In large pot, over medium-high heat, add rice and oil; stir to coat.

7. Add milk mixture; bring to a boil.
8. Reduce heat to medium-low and cover; cook 18 minutes.
9. Remove from heat; let stand covered 5 minutes.
10. Turn rice into large bowl; toss well to cool and separate.
11. Place in refrigerator, stirring often to help cool.
12. Note: If serving right away, toss with dressing to coat. If serving later, cover rice mixture and dressing separately.
13. Add desired amount of dressing right before serving.

Directions for dressing:

1. In food processor combine mango, onion, garlic, sugar, and red wine vinegar; purée until smooth.
2. While machine is running slowly, add oil in a very thin stream.
3. Add salt and pepper; pulse to combine.
4. To the rice, add peaches, red pepper, and cucumber.

One Cup Coconut Salad

This is an easy and delicious salad.

Ingredients:

1 c. mandarin oranges, drained
1 c. coconut
1 c. pineapple tidbits, drained
1 c. mini marshmallows
1 c. sour cream

Directions:

1. In large bowl, combine all ingredients.
2. Prepare several hours before serving and chill.

Fruit Salad with Papaya Mint Sauce

Use the freshest most flavorful fruit you can find, and you will enjoy this salad.

Ingredients for fruit salad:

½ lg. cantaloupe, peeled, seeded, cut into ½-inch pieces
1 med. papaya, peeled, seeded, cut into ½-inch pieces
½ lg. pineapple, peeled, cored, cut into ½-inch pieces
1 can whole litchi's, peeled, in heavy syrup, drained, halved lengthwise (11 oz.)
½ c. seedless green grapes, halved
½ c. seedless red grapes, halved
¼ c. coconut, sweetened, shredded or flaked, toasted
fresh mint sprigs

Ingredients for sauce:

1½ Tbs. fresh mint, coarsely chopped
1 lg. papaya, peeled, seeded, coarsely chopped
3 Tbs. fresh lime juice
3 Tbs. fructose sugar

Directions for fruit salad:

1. In large bowl, mix first 6 ingredients.
2. Cover; chill.
3. Can be made 4 hours ahead of time
4. To serve: Spoon fruit into 6 small bowls or goblets.
5. Drizzle sauce over fruit.
6. Sprinkle with coconut.
7. Garnish with mint.

Directions for sauce:

1. In blender or food processor, purée all ingredients until smooth; transfer to bowl.
2. Cover and refrigerate until ready to use.

3. Note: This can be prepared one day ahead.
4. Keep refrigerated.

Coconut Rice Salad

This makes an easy summer salad that tastes great on the next day!

Ingredients:

- 2¾ c. water
- ¼ c. sesame oil
- ¼ c. coconut, flaked, toasted
- ½ c. raisins
- ¼ c. slivered almonds, toasted
- 1 tsp. fish sauce
- 1 Tbs. curry paste
- 1 garlic clove, crushed
- 2 lg. limes, juiced
- 2 Tbs. peanut butter
- 2 c. basmati rice
- 1 can unsweetened coconut milk

Directions:

1. In saucepan, combine rice, coconut milk, and water.
2. Bring to boil, cover.
3. Reduce heat to low.
4. Simmer 15 to 20 minutes, until rice absorbs all liquid.
5. Set aside to cool.
6. In small bowl, stir together lime juice, peanut butter, oil, fish sauce, curry paste, and garlic.
7. Taste, and adjust flavor if needed.
8. When rice has cooled, stir in dressing.
9. Stir in coconut, raisins, and almonds.
10. Refrigerate 1 hour.
11. Serve with your favorite bread or buns.

French Beans and Coconut Salad

This dish is delicious on a hot summer day!

Ingredients:

1 coconut
1 lb. French beans
3 limes or 2 lemons
2 green chilies or 1 lg. piece ginger

Directions:

1. In medium saucepan, cook French beans in boiling water until tender.
2. Drain, rinse in cold water.
3. Put in bowl in the refrigerator.
4. Break coconut, grate flesh.
5. Mix with juice of limes or lemons.
6. Split chilies lengthwise and take out seeds, or peel ginger and chop fine.
7. Add to coconut; mix well.
8. Add mixture to French beans; mix well.
9. Cool in refrigerator 1 or 2 hours before serving.

Coconut Tropical Fruit Salad

This is a simple fresh fruit salad with a coconut milk dressing.

Ingredients for dressing:

1 banana, cut into pieces
1 orange, sectioned
⅓ c. coconut milk
3 tsp. orange marmalade

Ingredients for salad:

1 c. cantaloupe, made into balls

1 c. watermelon, made into balls
2 bananas, sliced
1 apple, cut into chunks
1 orange, peeled, cut into chunks
1 pear, cut into chunks
1 mango, cut into chunks
coconut flakes, unsweetened

Directions:

1. In large bowl, add all fruit.
2. In small bowl, combine dressing ingredients.
3. Blend until smooth.
4. Gently combine dressing with fruit.
5. Place into serving dishes; sprinkle with coconut flakes.
6. Note: Add a little orange juice if you want a thinner sauce.

Tropical Fruit Salad

This combination of fruit is very colorful and flavorful.

Ingredients:

1 mango, cut into chunks
2 kiwis, peeled, sliced
1 orange, peeled, cut into chunks
1 lg. banana, sliced
1 papaya, cut into chunks
⅓ c. coconut cream, thin with orange juice if thick

Directions:

1. In large bowl, add fruit.
2. Serve with your favorite dressing.
3. Drizzle 1 tablespoon of coconut cream over salad.
4. Serve with your favorite bread or buns.

Orange Fluff Salad

This is an easy-to-make salad that is very enjoyable.

Ingredients:

1 ctn. small curd cottage cheese (12 oz.)
1 pkg. orange gelatin (3 oz.)
1 can mandarin oranges, drained (11 oz.)
1 can pineapple, crushed, drained (15 oz.)
¾ c. coconut
½ c. pecans, chopped
¾ c. sweetened whipped cream

Directions:

1. In large mixing bowl, combine cottage cheese and gelatin; mix well.
2. Stir in oranges, pineapple, coconut, and pecans.
3. Fold in cream; refrigerate.

Fruit and Coconut Salad

Try this refreshing salad.

Ingredients:

1½ c. fresh pineapple, chopped
2 bananas, sliced
2 oranges, peeled, diced
2 apples, cored, diced
1 c. raisins
¾ c. coconut, shredded
¾ c. mayonnaise
lettuce leaves

Directions:

1. In medium bowl, combine all ingredients.
2. Serve on lettuce leaves.
3. Add one tablespoon of coconut oil per serving.

Side Dishes

Table of Contents

Page

Cabbage with Cinnamon and Coconut

Try this unique blend of flavors. It is actually very tasty.

Ingredients:

1 sm. head cabbage, chopped
1½ Tbs. olive oil
1 c. raisins
⅓ c. coconut, unsweetened, shredded
1 tsp. cinnamon

Directions:

1. In covered pan or skillet, heat all ingredients over medium-high heat.
2. Stir occasionally, 12 minutes or until cabbage is tender.

Cauliflower and Carrots with Coconut Dressing

This dish is a pleasant surprise combination of cauliflower and carrots with a delicious coconut dressing.

Ingredients for dressing:

¼ tsp. chili powder
¼ tsp. salt
1 garlic clove
4 tsp. lemon or lime juice
1 tsp. dark brown sugar
1 oz. coconut, unsweetened, desiccated, soaked in 4 Tbs. hot water for 30 minutes

Ingredients for vegetables:

1½ oz. sm. cauliflower florets

3 oz. carrots, cut into strips
¾ oz. salt

Directions:

1. In small bowl, combine first 5 dressing ingredients.
2. In large pot of boiling water, add salt.
3. Add cauliflower and carrots.
4. Boil rapidly for several minutes, until vegetables are tender but crispy.
5. Drain quickly.
6. Put in serving bowl.
7. Toss first with dressing, then with coconut.

Yields: 4 servings.

Coconut Sautéed Potatoes

Coconut and garlic flavor these sautéed potatoes.

Ingredients:

1 lg. baking potato, per person
2 garlic cloves, per person
1 Tbs. coconut oil, per person

Directions:

1. Preheat oven to 375 degrees.
2. Grease a baking pan with coconut oil.
3. Melt coconut oil on stove.
4. Slice potatoes thickly with skins on; place on prepared baking pan.
5. Crush garlic and spread over potato slices.
6. Pour melted coconut oil over potatoes and sprinkle sea salt over them.
7. Bake 20 minutes and turn.
8. Bake 20 minutes more, or until crunchy.

Coconut Carrots

The addition of coconut cream adds a nice taste for those of you who enjoy creamed carrots.

Ingredients:

2 lb. carrots, chopped
1 can cream of coconut (10 oz.)

Directions:

1. In medium saucepan, over medium heat, combine carrots and cream of coconut.
2. Simmer until carrots are tender, 20 minutes.

Yields: 8 servings.

Coconut Jasmine Rice

Coconut milk adds new qualities to rice. Enjoy!

Ingredients:

2½ c. jasmine ice
1¾ c. coconut milk
1¾ c. water
2 tsp. sugar
½ tsp. fine sea salt

Directions:

1. Rinse rice thoroughly, until water runs clear.
2. Drain well.
3. In saucepan, combine milk, water, sugar and salt.
4. Stir well, until sugar is dissolved.
5. Add rice and stir well.

6. Over medium-high heat, bring mixture to rolling boil.
7. Cover tightly, reduce heat.
8. Simmer 18 to 20 minutes, until all liquid absorbed.
9. Fluff and serve.

Red Beans and Greens in Coconut Milk

This recipe is excellent served over basmati or long-grain rice.

Ingredients:

1 c. red beans, soaked overnight
1 tsp. sea salt
1 onion, chopped
1 tomato, chopped
1 tsp. ground turmeric
1 bunch greens, kale, collard, or mustard, chopped
¼ c. coconut milk
cayenne pepper, to taste

Directions:

1. Drain and rinse beans.
2. In medium pot, add beans and enough water to cover.
3. Bring to boil.
4. Lower heat to medium, cook 30 minutes.
5. Add salt, onion, tomato, and turmeric.
6. Cook 30 minutes.
7. Add greens; cook 10 to 15 minutes until beans are soft.
8. Add milk and pepper.
9. Heat thoroughly.
10. Serve.

Coconut Sweet Potato Soufflé

Here is another great way to make sweet potatoes.

Ingredients:

3½ c. sweet potato, cooked, mashed
½ tsp. salt
½ c. butter, melted
½ c. half and half cream
1 c. white sugar
2 eggs
2 tsp. vanilla extract

Ingredients for topping:

⅓ c. flour
¾ c. butter, melted
1¼ c. coconut
1 c. brown sugar
1¼ c. pecans, chopped

Directions:

1. Preheat oven to 350 degrees F.
2. Lightly grease a soufflé dish.
3. In large bowl, add sweet potatoes.
4. Add remaining ingredients; beat well with mixer.
5. Pour contents into prepared dish.
6. Bake 15 minutes.
7. While this is baking, prepare topping.
8. In bowl, combine sugar, pecans, flour, and coconut.
9. Drizzle with melted butter; toss until well coated.
10. Remove from oven.
11. Sprinkle with topping.
12. Return to oven; bake 15 minutes until light brown.

Yields: 8 servings.

Curried Acorn Squash

Our family loves squash, and this is very flavorful.

Ingredients:

2½ lb. acorn squash
1½ Tbs. olive oil
1 tsp. black mustard seeds
1½ tsp. garlic, minced
1 tsp. ground coriander seeds
½ tsp. turmeric
1 tsp. salt
1 Tbs. brown sugar, lightly packed
3 Tbs. coconut, unsweetened, desiccated
2 c. water
1½ tsp. ground cumin
1 med. red onion, chopped

Directions:

1. Peel and seed squash; cut into 1½-inch pieces.
2. Heat a 4-quart heavy kettle over low heat until hot.
3. Cook onion in oil, stirring, until just softened.
4. Add mustard seeds and cook over moderate heat, stirring until they begin to pop.
5. Add garlic, cumin, coriander seeds, turmeric, salt, and brown sugar; cook while stirring, 30 seconds.
6. Add squash, stirring to coat with seasoning and add water.
7. Boil mixture over moderately high heat, stirring frequently for 10 minutes.
8. Add coconut and salt to taste and cook, stirring, until squash is tender and liquid is evaporated.

Yields: 4 servings.

Coconut Scented Rice with Almonds

This coconut flavored rice with almonds makes a nice side dish to your favorite main course.

Ingredients:

1 c. coconut milk
¼ tsp. salt
1 c. uncooked rice
⅓ c. almonds, sliced
¼ tsp. coconut extract

Directions:

1. In medium saucepan, over medium-high heat, bring milk and salt to boil.
2. Reduce heat, simmer, covered, 5 to 6 minutes, until liquid is absorbed.
3. Heat a large nonstick skillet over medium heat.
4. Dry-roast almonds until golden brown and fragrant, 1 to 5 minutes, stirring frequently.
5. Transfer to small plate, cool 1 minute.
6. Using back of fork or spoon, finely crush almonds.
7. Add extract to rice.
8. Fluff with fork; stir in almonds.

Yields: 4 to 6 servings.

Coconut Rice

The coconut milk adds an unexpected rich flavor to the rice.

Ingredients:

1½ c. canned unsweetened coconut milk, well stirred
½ c. water

1 tsp. salt
1 c. long-grain white rice
sweetened flaked coconut, toasted, for garnish

Directions:

1. In small saucepan, bring milk, water, and salt to a boil; stir in rice.
2. Reduce heat to low, cover and simmer for 15 minutes.
3. Remove pan from heat and let stand, covered, 5 minutes.
4. Serve rice sprinkled with coconut.

Coconut Squash Pie

This delightful dish tastes just like coconut pie.

Ingredients:

½ c. butter
1½ c. sugar
6 eggs
4 Tbs. flour
4 tsp. coconut extract
6 tsp. lemon extract
4 c. raw yellow squash, grated

Directions:

1. Preheat oven to 350 degrees F.
2. In large bowl, beat butter and sugar.
3. Add eggs, one at a time.
4. Add flour and extracts.
5. Divide squash into three pie shells.
6. Add butter mixture.
7. Sprinkle with coconut.
8. Bake 30 minutes.

Coconut Yam Bake

These yams are whipped for a delightful side dish.

Ingredients for casserole:

¼ c. cream
3 med. fresh yams
1 c. sugar
½ c. butter
1 c. coconut, shredded
2 lg. eggs, beaten
1 tsp. vanilla extract

Ingredients for topping:

1 c. brown sugar
1 c. flour
½ c. butter, melted
1 c. pecans, chopped

Directions for casserole:

1. Preheat oven to 325 degrees F.
2. Wash, peel, and cut yams into cubes.
3. In saucepan, add yams, cover with water.
4. Cook until tender.
5. Drain water.
6. Mash yams to make 3 cups.
7. Add sugar, butter, coconut, milk, eggs, and vanilla.
8. Beat until creamy.
9. Pour into a 2 or 3-quart casserole dish.
10. Set aside.

Directions for topping:

1. In small bowl, mix sugar and flour until smooth.
2. Add butter; mix well.

3. Add pecans.
4. Spread or crumble the mixture on top of the prepared casserole.
5. Bake 30 minutes.

Yields: 4 to 6 servings.

Garlic Ginger Coconut Jasmine Rice

Garlic, ginger, and coconut are an excellent accompaniment to rice. Enjoy!

Ingredients:

- 2 c. white, raw jasmine rice
- 2 Tbs. olive oil
- 4 garlic cloves, crushed
- 1 Tbs. fresh ginger, grated
- ¼ c. coconut milk
- 2½ c. water
- salt, to taste

Directions:

1. In saucepan, over medium heat, sauté oil, garlic, and ginger briefly.
2. Add rice and stir well.
3. Add milk, water, and salt; bring to boil.
4. Give a quick stir, cover.
5. Reduce heat to low; cook 15 minutes.
6. Remove from heat.
7. Let stand, covered, five minutes.
8. Fluff with fork.

Yields: 4 servings.

Sweet Potato Coconut S'mores

Our children love s'mores, and this adds a delightful addition for a sweet surprise.

Ingredients:

1 lg. sweet potato, peeled, sliced into 6¼-inch slabs
16 graham crackers
1 egg beaten with 1Tbs. water
8 oz. milk or dark chocolate bar, chopped into pieces
½ c. butter 6 jumbo toasted coconut marshmallows
flour, for dusting

Directions:

1. Preheat oven to 450 degrees F.
2. In steamer, bring water to boil with a dash of salt.
3. Trim sweet potato into roughly 2 x 4-inch rectangles.
4. Place them in steamer; cover.
5. Cook until just tender, 5 to 7 minutes.
6. Remove cooked slices to a rack covered with paper towels; cool to room temperature.
7. Crush graham crackers in plastic bag with rolling pin.
8. In three separate shallow bowls, place flour, egg, and cracker crumbs.
9. One at a time, dip each slice of sweet potato in flour, patting off excess.
10. Dip in egg and coat with crumbs.
11. Place coated slices on platter in freezer, 10 minutes.
12. Place chocolate in double boiler above gently simmering water to melt.
13. Heat a large nonstick skillet over medium heat.
14. Melt butter, fry graham coated sweet potato or yam slices for 2 to 3 minutes per side.

15. Place marshmallow on top of each and put the whole skillet in hot oven, covered.
16. Bake 4 to 6 minutes, until marshmallows are golden and puffed.
17. Remove from oven; place each s'more on a plate.
18. Drizzle with melted chocolate and serve immediately.

Yields: 6 servings.

Coconut Wild Rice

This is a very flavorful way to cook wild rice.

Ingredients:

2 c. wild rice
2 c. water
2½ c. coconut milk
½ tsp. sea salt
coconut flakes, for garnish

Directions:

1. In cold water, wash rice several times until clear.
2. In heavy-bottomed saucepan, place all ingredients, except coconut flakes.
3. Bring to boil.
4. Cover; reduce heat to low.
5. Simmer until all water is absorbed and rice is just beginning to stick to bottom of pot.
6. Turn off heat.
7. Allow rice to rest 5 to 10 minutes.
8. Gently fluff rice before serving.
9. Garnish with coconut flakes.

Lemongrass Ginger Noodles

Noodles are another popular classic. Try these spicy ginger noodles for a tasty vegetarian dish.

Ingredients:

4 c. lemongrass tea
1 can coconut milk
1 lg. sweet onion, chopped
2 Tbs. peanut butter
dash cayenne
Chinese noodles
fresh gingerroot
salt
snow peas or broccoli
sesame oil

Directions:

1. Brew tea 5 min; set aside.
2. Cook noodles per package instructions; set aside.
3. Steam vegetables, until bright in color.
4. In frying pan, sauté onions in oil.
5. Add half of coconut milk, cook until semi-soft.
6. Add peanut butter, stir to dissolve.
7. Grate ginger to taste.
8. Add remainder of milk.
9. Add cayenne and salt to taste.
10. Add ½ tsp. sesame oil.
11. Stir in noodles and vegetables.
12. Variations: Add tofu, chopped in cubes. Use lemon juice in addition to lemongrass tea for added flavor.

Yields: 2 servings.

Coconut Delights Cookbook

A Collection of Coconut Recipes
Cookbook Delights Series – Book 4

Soups

Table of Contents

Page

Thai Coconut Soup

This soup is easy to make and makes the perfect comfort food. You can also substitute shrimp for chicken. Enjoy!

Ingredients:

¼ c. lime juice
¼ c. brown sugar
1 lb. med. shrimp, peeled, deveined
2 cans coconut milk (13.5 oz.)
2 c. water
1 pc. galanga, 1-inch thin slice
10 kaffir lime leaves, torn in half
1 lb. shiitake mushrooms, sliced
1 tsp. curry powder
1 Tbs. green onion, thinly sliced
1 tsp. dried red pepper flakes
3 Tbs. fish sauce
4 stalks lemongrass, bruised, chopped

Directions:

1. Bring a pot of water to boil.
2. Boil shrimp until cooked, one minute.
3. Drain and set aside.
4. Pour coconut milk and 2 cups of water in large saucepan; bring to a simmer.
5. Add galanga, lemongrass, and lime leaves; simmer 10 minutes.
6. Strain coconut milk into a new pan, discard spices.
7. Simmer mushrooms in coconut milk for 5 minutes.
8. Stir in lime juice, fish sauce, and brown sugar.
9. Season to taste with curry powder.
10. Reheat shrimp in the soup.
11. Serve.
12. Garnish with green onion and red pepper flakes.

Coconut Asparagus Soup

Fresh spring asparagus makes a delicious soup, and the coconut milk adds great flavor.

Ingredients:

1 lb. fresh asparagus, trimmed, cut into 1-inch pieces
½ c. onion, chopped
½ c. celery, chopped
½ c. chicken broth
4 Tbs. coconut oil
6 Tbs. flour
1 tsp. sea salt
1 pinch cayenne pepper
1 tsp. marjoram fresh or 1 Tbs. dried
2 cans coconut milk (13.5 oz.)
1 tsp. lemon juice, freshly squeezed

Directions:

1. In large saucepan, combine asparagus, onion, and chicken broth.
2. Cover; over high heat, bring to boil.
3. Reduce heat immediately.
4. Uncover, simmer 15 minutes.
5. Take half this mixture and purée in a blender.
6. Pour other half of mixture into bowl; set aside.
7. In same saucepan, over medium heat, melt coconut oil.
8. Stir in flour and whisk.
9. Cook, stirring constantly, until mixture turns golden brown being careful not to burn.
10. Add spices; reduce heat to low.
11. Stir in puréed vegetables and cooked vegetables.
12. Add lemon juice, and milk.
13. Stir frequently, do not boil.

Chicken Coconut Vegetable Soup

This is a delicious and flavorful soup.

Ingredients:

½ c. snow peas, julienned
½ c. carrots, julienned
4 stalks fresh lemongrass
4 c. chicken stock
15 pc. gingerroot, quarter-size slices
10 whole black peppercorns, crushed
3 serrano chiles, thinly sliced
12 fresh lime leaves or ¼ tsp. grated lime zest
2 c. coconut milk
1 chicken breast, boneless, skinless, cut into pieces
1 c. canned straw mushrooms, drained
2 Tbs. fish sauce
juice of 2 limes
fresh cilantro leaves, for garnish

Directions:

1. Remove grassy tops of lemongrass, leaving stalks 6 inches long, cutting off any hard root section.
2. Using blunt edge of knife or cleaver, bruise each stalk all over.
3. Cut each stalk into 4 pieces.
4. In large saucepan, bring chicken stock to a boil, reduce to simmer.
5. Add lemongrass, gingerroot, peppercorns, serrano chile, and lime leaves.
6. Simmer 15 minutes.
7. Add coconut milk and cook 5 minutes.
8. Add chicken pieces, mushrooms, snow peas, and carrots.
9. Simmer 10 to 12 minutes.
10. Stir in fish sauce and lime juice.

11. Serve immediately.
12. Garnish with cilantro leaves.
13. Note: Be careful to avoid chewing the lemongrass, galanga, ginger, or lime leaves when eating.

Chicken Coconut Soup

Our family loves this soup, and it is very easy to make. Company always enjoys it also.

Ingredients:

¾ lb. boneless, skinless chicken
3 Tbs. olive oil
2 c. water
2 Tbs. fresh gingerroot, minced
4 Tbs. fish sauce
¼ c. fresh lime juice
¼ Tbs. cayenne pepper
½ tsp. ground turmeric
2 Tbs. green onion, thinly sliced
1 Tbs. fresh cilantro, chopped
2 cans coconut milk (14 oz.)

Directions:

1. Cut chicken into thin strips.
2. Sauté in oil, 2 to 3 minutes.
3. In pot, bring coconut milk and water to boil.
4. Reduce heat.
5. Add gingerroot, fish sauce, lime juice, pepper, and turmeric.
6. Simmer 10 to 15 minutes, until chicken is done.
7. Sprinkle with scallions and cilantro.
8. Serve steaming hot with your favorite fresh bread or buns.

Coconut Tuna Chowder

Try this tuna chowder with coconut cream. It makes an interesting flavor combination.

Ingredients:

3 cans tuna (7 oz.)
¼ c. virgin coconut oil
⅓ c. coconut flakes, toasted
3 Tbs. flour
2 c. chicken broth
½ tsp. salt
1 lg. onion, sliced
1 tsp. curry powder
3 celery stalks, chopped
¼ tsp. ginger threads
2 whole cloves
2 c. milk
6 peppercorns
½ c. coconut cream
½ c. half and half cream
1 bay leaf

Directions:

1. Drain and flake tuna.
2. In oven, toast coconut flakes until golden brown, 5 to 8 minutes.
3. In large pot, add broth, onion, celery, cloves, bay leaf, and peppercorn.
4. Simmer 30 minutes.
5. In small bowl, mix flour, salt, curry powder, and ginger.
6. Add a small amount of cream to flour mixture to make a paste.
7. In large skillet, add oil.
8. Slowly add milk; whisk in flour mixture, mix well.

9. Bring to boil, stirring constantly until sauce thickens.
10. Add broth mixture, removing bay leaf.
11. Add tuna and cream.
12. Ladle into serving bowls, top with coconut flakes.

Spinach and Coconut Soup

This soup is very tasty.

Ingredients:

1 lb. spinach
1 lg. onion, finely chopped
1 lg. potato, finely diced
4 Tbs. butter
2 garlic cloves, finely chopped
4 oz. creamed coconut
3 c. vegetable stock
¼ pt. half and half cream
black pepper, to taste

Directions:

1. In heavy saucepan, melt butter.
2. Add onions and garlic; cook gently, until transparent.
3. Add potato, cook 5 minutes.
4. Add spinach and vegetable stock.
5. Simmer until potatoes are cooked.
6. Pour into blender container, blend until smooth.
7. Return to saucepan, over low heat, add creamed coconut.
8. Stir until melted; add cream.
9. Add freshly ground pepper, to taste.
10. Ladle into soup bowls and serve immediately with your favorite bread or buns.

Pumpkin and Coconut Soup

Traditional pumpkin and coconut soup is usually quite hot and spicy.

Ingredients:

1 Tbs. coconut oil
1 Tbs. butter
1 garlic clove, chopped
4 shallots, chopped
2 sm. red chili peppers, chopped
1 Tbs. lemongrass, chopped
2⅛ c. chicken stock
1½ c. unsweetened coconut milk
1 bunch fresh basil leaves

Directions:

1. In medium saucepan, heat butter over low heat.
2. Sauté garlic, shallots, chilies, and lemongrass in oil until fragrant. Be careful not to burn the garlic.
3. Stir in chicken stock, coconut milk, and pumpkin and bring almost to a boil.
4. Simmer on low heat until pumpkin softens.
5. In a blender, blend the soup to the consistency you prefer, smooth or chunky.
6. Pour back into the soup pan; heat and serve with basil leaves.
7. Season to taste with your choice of seasonings.

Indian Coconut Soup

This is quick, and easy to make and has a nice texture.

Ingredients:

2 Tbs. butter
½ c. coconut, shredded
¼ c. pistachios, chopped

3 Tbs. honey
½ tsp. cardamom
½ tsp. cinnamon
½ tsp. nutmeg
¼ tsp. fennel
2 c. milk
2 cans coconut milk, well shaken

Directions:

1. In medium pot, sauté coconut in butter until toasted, stirring well.
2. Add pistachios, sautéing a little longer.
3. Add honey and spices; stir.
4. Add both milks; stirring slowly.
5. Heat, but do not boil.
6. Serve garnished with pistachios.

Crabmeat Soup

Serve this easy-to-make, delicious soup with French bread.

Ingredients:

2 cans asparagus soup
2 cans coconut cream
1 Tbs. curry
½ can water
5½ oz. fresh crabmeat
1 Tbs. soy sauce
squeeze of lemon juice
salt and pepper

Directions:

1. In medium saucepan, mix all ingredients together except the crabmeat; heat through.
2. Add crabmeat and heat again.

Coconut Vegetable Soup

This is a delicious vegetarian soup.

Ingredients:

- 2 shallots, minced
- 2 garlic cloves, minced
- 1 Tbs. gingerroot, minced
- ¼ tsp. red pepper flakes, crushed
- 2 Tbs. peanut oil
- 1 carrot, julienned
- 1 med. zucchini, julienned
- 1½ c. napa cabbage, shredded
- 2½ c. fresh bean sprouts
- 1 celery rib, finely sliced
- 4 c. chicken stock
- 1 c. unsweetened coconut milk
- 1 stalk lemongrass, trimmed, pounded, chopped
- ½ tsp. turmeric
- salt, to taste
- fried red onion flakes, for garnish
- lemon wedges, for garnish

Directions:

1. In large saucepan, heat oil.
2. Add shallots, garlic, ginger, and crushed red peppers; cook until aromatic.
3. Add carrots, zucchini, cabbage, and celery, stirring to coat vegetables with spices.
4. Add stock, milk, lemongrass, and turmeric.
5. Simmer, uncovered, on low heat, 15 to 20 minutes until vegetables are tender.
6. To serve, place a small handful of bean sprouts in bottom of each bowl and add soup.
7. Garnish with onion flakes.
8. Serve with a squeeze of lemon wedge.

Spicy Sweet Potato and Coconut Soup

This is a very flavorful soup.

Ingredients:

3 lg. sweet potatoes, rinsed
1 Tbs. olive oil
1 onion, chopped
4 Tbs. fresh ginger, peeled, thinly sliced
1 Tbs. red curry paste
1 can unsweetened coconut milk (15 oz.)
3 c. chicken or vegetable broth
3½ Tbs. lemon juice
1 tsp. kosher salt
1 Tbs. toasted sesame oil
½ c. fresh cilantro stems

Directions:

1. Preheat oven to 400 degrees F.
2. Put sweet potatoes on oven rack.
3. Bake 50 minutes, until tender.
4. Remove from oven; cool.
5. In large saucepan, over medium heat, add olive oil.
6. Sauté onion and ginger until soft, 5 minutes.
7. Stir in curry paste, cook 1 minute.
8. Add milk and broth; gently bring to boil.
9. Reduce heat to simmer, partially cover; cook 5 minutes.
10. Skin potatoes; cut into 1-inch chunks.
11. Add potatoes to other mixture, simmer 5 minutes.
12. Stir in lemon juice and salt.
13. Ladle soup into serving bowls.
14. Drizzle sesame oil evenly over soup.
15. Garnish with the cilantro.

Yields: 4 servings.

Squash Coconut Milk Soup

This soup is very flavorful with a delicious blend of coconut and squash.

Ingredients:

4 c. butternut squash
2 c. spinach leaves
1 can coconut milk (14 oz.)
1 c. water
1 c. onion, thinly sliced
1 Tbs. fish sauce or soy sauce
1 tsp. chili paste
1 tsp. brown sugar
cilantro, for garnish
limes, for garnish

Directions:

1. Preheat oven to 400 degrees F.
2. Cut butternut squash in half lengthwise.
3. Bake face down, until soft.
4. Wash spinach leaves.
5. In large pot or Dutch oven, heat milk.
6. Add water, onion, fish sauce, paste, and sugar.
7. Simmer on low heat.
8. When squash is done, remove seeds and scoop out 2 to 4 cups flesh. Reserve remainder.
9. Add squash to soup mixture.
10. Mash with potato masher until well mixed.
11. Cook 10 minutes.
12. Chop spinach leaves.
13. Place into individual bowls.
14. Ladle the soup on top.
15. Chop cilantro leaves.
16. Cut limes or lemons into quarters.
17. Serve with the soup.

Coconut Bean Soup

This is a spicy soup that is quick and easy-to-make. Serve with a fresh salad, and your favorite bread or buns.

Ingredients:

3 Tbs. butter
1 onion, diced
1 green chili, seeded, finely diced
2 garlic cloves, crushed
1 red pepper, seeded, diced
1 tsp. curry powder
2 tomatoes, skinned, diced
1 can red kidney beans, drained
1 can coconut milk
1 c. water or chicken stock
1 c. chicken, cooked, finely chopped
¾ c. rice, cooked
3 Tbs. desiccated coconut
2 Tbs. fresh coriander

Directions:

1. In large pot, melt butter.
2. Sauté onion, pepper, chili, and garlic until soft.
3. Add tomatoes, milk, and water or stock.
4. On low heat, simmer 10 minutes.
5. Add chicken and rice.
6. Simmer until heated through, 2 minutes.
7. Season to taste.
8. Ladle into soup bowls.
9. Garnish with coconut and coriander.
10. Serve immediately.

Yields: 6 servings

Coconut and Seafood Soup with Garlic Chives

This recipe makes a delicious soup. The garlic chives add great flavor.

Ingredients:

2½ c. fish stock
1 Tbs. olive oil
1 bunch garlic chives, chopped
½ oz. fresh coriander (cilantro)
3 Tbs. Thai fish sauce
4 Tbs. Thai green curry paste
2 lemongrass stalks, chopped
3 kaffir lime leaves, shredded
5 pc. fresh gingerroot, cut into thin slices
4 shallots, chopped
1 can coconut milk (14 oz.)
1 lb. uncooked jumbo shrimp, peeled, deveined
1 lb. prepared squid
4 Tbs. fried shallot slices, to serve
lime juice (optional)
salt and ground black pepper, to taste

Directions:

1. Pour stock into a pan; add gingerroot, lemongrass, half of the lime leaves, half of the chives, and coriander stalks.
2. Bring to a boil; then simmer for 20 minutes; strain.
3. Rinse pan; add oil and shallots.
4. Over medium heat, cook 5 to 10 minutes, stirring occasionally, until light brown.
5. Stir in stock, coconut milk, remaining lime leaves and half the fish sauce.
6. Heat gently until soup is simmering; cook over low heat, 5 to 10 minutes.

7. Stir in curry paste and peeled prawns; cook 3 minutes.
8. Add squid; cook 2 more minutes.
9. Add lime juice, if desired.
10. Season to taste.
11. Stir in remaining fish sauce, chives, and coriander leaves.
12. Serve in warm, shallow bowls sprinkled with fried shallots.
13. Note: Instead of squid, add 14 oz. of firm white fish, such as monkfish or fresh mussels.
14. Steam 1½ pounds closed mussels in a tightly covered pan for 3 minutes, or until shells have opened. Discard any that remain shut.
15. Remove mussels from their shells.

Red Pepper Curry Coconut Soup

This is a delicious spicy soup to try.

Ingredients:

3 red bell peppers, chopped
2 sm. onions, diced
1 Tbs. butter
1 Tsp. curry paste
1 c. coconut milk
3 c. chicken or vegetable stock

Directions:

1. In medium saucepan, melt butter.
2. Add onions, sauté until translucent.
3. Add curry paste, fry 3 minutes.
4. Add stock and peppers; simmer 20 minutes.
5. In blender, purée in batches.
6. Return soup to pan.
7. Stir in milk, reheat and serve immediately.

Coconut Lobster Prawn Soup

This soup is a welcome treat for shrimp lovers.

Ingredients for prawn stock:

½ tsp. dried thyme
½ oz. butter
½ c. dry vermouth
½ sm. red onion, peeled, chopped
1 stalk of celery, chopped
1 carrot, peeled, chopped
3 lb. prawns, peeled, heads and shells reserved
6 c. cold water
2 bay leaves
freshly ground pepper, to taste

Ingredients for soup:

1 Tbs. butter
5 oz. smoked bacon, finely sliced
½ sm. red onion, peeled, chopped
8 oz. fresh lobster mushrooms, chopped
1 sm. red bell pepper, seeded, chopped
1 sm. yellow bell pepper, seeded, chopped
1 lg. ripe tomato, diced
1 lb. acorn squash, seeded, peeled, diced
4 lg. garlic cloves, chopped
¼ tsp. cayenne pepper
½ tsp. salt
3 Tbs. curry powder
4 c. chicken stock
1 can coconut milk (12 oz.)
1 can kernel corn, rinsed (12 oz.)
2 med. potatoes, peeled, diced
½ c. heavy cream
2 Tbs. cornstarch, dissolved in same amount of water
2 Tbs. parsley, finely chopped

Directions for prawn stock:

1. In small heavy pot, melt butter.
2. Over low heat, cook onion, celery, carrot, and shrimp, 3 minutes.
3. Add vermouth, water, bay leaves, thyme, and pepper.
4. Bring to boil.
5. Simmer, 30 minutes.

Directions for soup:

1. In large heavy pot, melt butter.
2. On medium heat, add bacon, cook 3 minutes.
3. Add vegetables, except corn and potatoes, garlic cloves, pepper, and curry powder.
4. Reduce heat to low, cook 5 minutes, stirring occasionally.
5. Add chicken stock.
6. Strain over prawn stock.
7. Add milk, corn, salt, and pepper; simmer 10 minutes.
8. Add diced potatoes, simmer 30 minutes.
9. In small cup, combine cream and cornstarch dissolved in water; bring to boil.
10. Add shrimp, cook 2 minutes.
11. Serve in warm soup bowl.
12. Garnish with parsley.
13. Serve with your favorite bread or buns.

Yields: 8 servings.

Did You Know? . . . .

Did you know that the oil from your skin as well as water will darken coconut jewelry also?

Coconut Lentil Soup

Serve this lovely, smooth soup with croutons for a nice change of pace.

Ingredients:

- 6 oz. red lentils
- 1 pc. fresh ginger, ¾-inch
- 1 onion, finely chopped
- 1 garlic clove, finely chopped
- 1 Tbs. olive oil
- 1 tsp. turmeric
- 1¾ c. chicken stock
- 1 can crushed tomatoes (16 oz.)
- ¾ c. coconut milk

Directions:

1. In colander rinse lentils, drain.
2. In medium pot, heat oil.
3. Sauté ginger, onion, and garlic until soft.
4. Add turmeric, chicken stock, tomatoes, and lentils.
5. Bring to boil, simmer 20 minutes.
6. In blender, purée soup until smooth.
7. Return to pot.
8. Add milk, heat over low heat.
9. Ladle into soup bowls and serve immediately with your favorite croutons, bread, or buns.

Yields: 4 servings.

Did You Know? . . . .

Did you know that coconut oil was the world's leading vegetable oil until soybean oil took over in the 1960s?

Coconut Delights Cookbook
A Collection of Coconut Recipes
Cookbook Delights Series – Book 4

Wines and Spirits

Table of Contents

About Cooking with Alcohol

Some recipes in this cookbook contain, among other ingredients, liquors. It is for the purpose of obtaining desired flavor and achieving culinary appreciation and not to be abused in any way. In cooking and baking, alcohol evaporates and only the flavor may be enjoyed. When mixed in cold, however, such as in desserts, caution must be exercised. These recipes are intended for people who may consume small amounts of alcohol in a responsible and safe manner.

I live in Washington State and we are proud of our wine production. Washington State is rapidly gaining prestige as a premier wine producer. Do enjoy the art of wine tasting and enjoy the completeness and uniqueness of each wine. It is an art to enjoy and savor in moderation.

If consumption of even small amounts of alcoholic ingredients presents a problem, in whatever form, please substitute coffee flavor syrups, found in coffee sections of supermarkets. For example, instead of Southern Comfort liqueur, substitute with Irish Cream or Amaretto Syrup.

Karen Jean Matsko Hood

Coconut Loco

This makes a pleasing mixed drink. Enjoy!

Ingredients:

1 can cream of coconut
8 oz. **each:** grenadine, tequila, vodka, gin, rum
crushed ice

Directions:

1. In blender, add all ingredients with crushed ice.
2. Pour into chilled glasses and serve.

Coconut Rum Blend

This makes a great rum blend.

Ingredients:

1 oz. coconut rum
½ oz. amaretto
4 oz. orange juice
½ oz. grenadine

Directions:

1. Combine all ingredients.
2. Add ice.
3. Serve.

Mango and Coconut Daiquiri

This is a delicious daiquiri.

Ingredients:

1 c. mango, largely diced
1 c. ice
2 oz. white rum
2 Tbs. fresh lime juice
2 Tbs. coconut cream
1 tsp. sugar
coconut shavings, for garnish

Directions:

1. In a blender, mix all ingredients except coconut shavings.
2. Blend until smooth.
3. Garnish with coconut shavings.
4. Pour into chilled glasses.

Coconut Margarita

Margaritas are always popular. Try this coconut version.

Ingredients:

1 c. ice
1 tsp. cream of coconut
2 oz. pineapple
2 oz. tequila
2 oz. sweet and sour
½ oz. orange-flavored liqueur
1 tsp. shaved, toasted coconut, for rim

Directions:

1. In blender, place ice.
2. Add remaining ingredients except toasted coconut.
3. Blend for 8 seconds.
4. To make drink thicker, add more ice and blend.
5. Dip wet rim of a large margarita glass in toasted coconut.
6. Pour blended mixture in the glass.

Yields: 1 drink.

Coconut Pineapple Punch

This is a unique punch that is soothing.

Ingredients:

½ c. vodka
1 can coconut cream (15 oz.)
3 c. canned pineapple juice
1 Tbs. powdered sugar
2 Tbs. Crème de Menthe

Directions:

1. In large bowl, combine coconut cream, vodka, pineapple juice, and powdered sugar.
2. Mix well; stir in crème de menthe.
3. Refrigerate before serving.

Coconut Snowball Cocoa

This makes an attractive presentation for serving a hot, soothing beverage. Enjoy!

Ingredients:

½ c. unsweetened cocoa
½ c. chocolate ice cream sauce
½ c. dark rum (optional)
1 qt. milk
1 pt. vanilla ice cream
1 c. coconut, flaked
1 c. cream of coconut
1 tsp. coconut extract
8 maraschino cherries

Directions:

1. Scoop ice cream into 8 small balls.
2. Immediately roll in coconut.
3. Place on wax paper-lined baking sheets.
4. Freeze until ready to use.
5. In large saucepan, whisk cocoa into milk.
6. Stir in rum, if desired, cream of coconut, and coconut extract.
7. Over medium-high heat, bring to simmer.
8. Pour into 8 large heatproof mugs.
9. Float ice cream balls in cocoa.
10. If desired, drizzle ice cream ball with chocolate sauce and top with a cherry.

Coconut Breeze

One sip of this and you will be able to feel the tropical breeze. This is a very smooth and flavorful drink.

Ingredients:

1½ oz. coconut rum
2 oz. cranberry juice
2 oz. pineapple juice

Directions:

1. Combine all ingredients.
2. Add ice.
3. Serve.

Coconut Cloud Coffee

This flavored coffee goes well with dessert.

Ingredients:

1⅓ c. coconut, flaked (3½ oz. can)
¼ c. rum extract
2 c. milk
2 c. hot, strong, brewed coffee
whipped cream, for garnish
toasted coconut, for garnish
almonds, toasted, finely chopped, for garnish

Directions:

1. In saucepan, combine milk and coconut.
2. Heat over medium-low heat, stirring occasionally, until mixture steams.
3. In blender, pour hot milk mixture into container.
4. Cover, blend on high speed until smooth.

5. Strain.
6. In same saucepan, combine strained milk and hot coffee.
7. Heat thoroughly but do not boil.
8. Remove from heat.
9. Stir in rum extract.
10. Pour into 4 glasses or coffee cups.
11. Top each with a dollop of whipped cream.
12. Sprinkle toasted coconut and toasted almonds over each serving.
13. Serve immediately.

Yields: 4 servings.

Pineapple Coconut Cooler

This blender beverage is a refreshing new version of the pina colada.

Ingredients:

- 2½ tsp. coconut extract
- ½ tsp. vanilla extract
- 1 tsp. rum extract
- 2 c. cold pineapple juice
- 2 c. ice cubes
- 2 c. pineapple sherbet

Directions:

1. In blender container, place all ingredients, except sherbet; cover.
2. Blend on high speed until ice is crushed.
3. Add sherbet; cover; blend until smooth.
4. Garnish with pineapple slices and cherries, if desired.
5. Pour into chilled glasses.

Coconut Canal

Try this coconut cocktail.

Ingredients:

¼ can of condensed milk
2 coconuts and all water
1 Tbs. gin
1 tsp. aromatic bitters

Directions:

1. Mix coconut water and jelly with condensed milk.
2. Add gin and bitters.
3. Chill slightly.
4. Serve with cracked ice.

Coconut Grove Drink

Coconut rum provides a nice flavor combination for this tropical drink.

Ingredients:

1½ oz. coconut rum
1 oz. Crème de Bananes
1 oz. white rum
4 oz. pineapple juice
1 tsp. lemon juice

Directions:

1. Mix all ingredients together.
2. Shake and strain into ice-filled glass.
3. Garnish with pineapple and lemon slices.

Ambrosia Punch

Try this refreshing punch on a hot summer day.

Ingredients:

1 can pineapple, crushed, undrained (20 oz.)
1 can cream of coconut (15 oz.)
2 c. apricot nectar, chilled
2 c. orange juice, chilled
1½ c. light rum
1 lg. bottle club soda, chilled

Directions:

1. In blender, purée pineapple and cream of coconut until smooth.
2. In a punch bowl, combine the puréed mixture, apricot nectar, orange juice, and rum. Mix well.
3. Just before serving, add club soda and serve over ice.

Lava Flow

This is a very refreshing drink with many fruit flavors.

Ingredients:

1 oz. light rum
1 oz. dark rum
2 oz. strawberries
1 banana, ripe
2 oz. pineapple juice, unsweetened
2 oz. coconut cream

Directions:

1. In a blender, purée strawberries and both rums until a smooth paste forms; pour into a hurricane glass.
2. Rinse the blender; add banana, coconut cream, and pineapple juice. Blend until smooth.
3. Pour banana mixture into strawberry paste slowly.

Coconut Pina Colada

Try this refreshing coconut pina colada.

Ingredients:

1½ oz. light rum
1½ oz. coconut cream
1½ oz. pineapple, crushed
4 oz. crushed ice
orange slice, pineapple wedge, cherry, for garnish
whipped topping (optional)

Directions:

1. In a blender, add rum, coconut cream, crushed pineapple, and ice; blend on high until smooth.
2. Pour into chilled cocktail glass.
3. Top with whipped cream, if desired.
4. Garnish with orange slice, pineapple wedge, and cherry.

Blue Hawaiian

Try this refreshing coconut drink on a summer day.

Ingredients:

1 oz. light rum
½ oz. Blue Curacao
1 oz. pineapple juice
1 oz. cream of coconut
1 tsp. sugar
strawberries or pineapple slices, for garnish

Directions:

1. Mix all ingredients in a shaker jar half filled with ice cubes.
2. Strain into a large glass half filled with crushed ice.
3. Decorate with a paper umbrella and strawberries or a slice of pineapple.

Festival Information

Hainan (Coconut Island) International Coconut Festival
Date: The first 10-day period in April (the third day of the third lunar month).
Place: Haikou City, Wenchang County, Tongzha City, and Sanya City in Hainan Province in China.

Kauai Coconut Festival
Date: October
Place: Kapaa Beach Park, Kauai, Hawaii

Pesta Kelapa (Coconut Fest)
Date: September
Place: Sabah, Malaysia in the Northern district of Kudat.

San Pablo City, Laguna - Philippines Coconut Festival
Date: January
Place: 87 kms. south of Manila

Coconut Associations and Commissions

United Coconut Associations of the Philippines, Inc.
(UCAP)
2nd Flr., PCRDF Bldg.,
Pearl Drive cor. Lourdes St., Ortigas Center, 1605 Pasig City, Metro Manila, Philippines.
Tel. Nos. (63-02) 633-80-29 or (63-02) 633-9286
Fax. No. (63-02) 633-8030

U.S. and Metric Measurement Charts

Here are some measurement equivalents to help you with exchanges. There was a time when many people thought the entire world would convert to the metric scale. While most of the world has, America still has not. Metric conversions in cooking are vitally important to preparing a tasty recipe. Here are simple conversion tables that should come in handy.

U.S. Measurement Equivalents

a few grains/pinch/dash (dry) = less than ⅛ teaspoon
a dash (liquid) = a few drops
3 teaspoons = 1 tablespoon
½ tablespoon = 1½ teaspoons
1 tablespoon = 3 teaspoons
2 tablespoons = 1 fluid ounce
4 tablespoons = ¼ cup
5⅓ tablespoons = ⅓ cup
8 tablespoons = ½ cup
8 tablespoons = 4 fluid ounces
10⅔ tablespoons = ⅔ cup
12 tablespoons = ¾ cup
16 tablespoons = 1 cup
16 tablespoons = 8 fluid ounces
⅛ cup = 2 tablespoons
¼ cup = 4 tablespoons
¼ cup = 2 fluid ounces
⅓ cup = 5 tablespoons plus 1 teaspoon
½ cup = 8 tablespoons
1 cup = 16 tablespoons
1 cup = 8 fluid ounces
1 cup = ½ pint
2 cups = 1 pint
2 pints = 1 quart
4 quarts (liquid) = 1 gallon
8 quarts (dry) = 1 peck
4 pecks (dry) = 1 bushel
1 kilogram = approximately 2 pounds
1 liter=approximately 4 cups or 1quart

Approximate Metric Equivalents by Volume

U.S.	Metric
¼ cup = 60 milliliters	
½ cup = 120 milliliters	
1 cup = 230 milliliters	
1¼ cups = 300 milliliters	
1½ cups = 360 milliliters	
2 cups = 460 milliliters	
2½ cups = 600 milliliters	
3 cups = 700 milliliters	
4 cups (1 quart) = .95 liter	
1.06 quarts = 1 liter	
4 quarts (1 gallon) = 3.8 liters	

Approximate Metric Equivalents by Weight

U.S.	Metric
¼ ounce = 7 grams	
½ ounce = 14 grams	
1 ounce = 28 grams	
1¼ ounces = 35 grams	
1½ ounces = 40 grams	
2½ ounces = 70 grams	
4 ounces = 112 grams	
5 ounces = 140 grams	
8 ounces = 228 grams	
10 ounces = 280 grams	
15 ounces = 425 grams	
16 ounces (1 pound) = 454 grams	

Glossary

Aerate: A synonym for sift; to pass ingredients through a fine-mesh device to break up large pieces and incorporate air into ingredients to make them lighter.

Al dente: "To the tooth," in Italian. The pasta is cooked just enough to maintain a firm, chewy texture.

Baste: To brush or spoon liquid fat or juices over meat during roasting to add flavor and prevent drying out.

Bias-slice**:** To slice a food crosswise at a 45-degree angle.

Bind: To thicken a sauce or hot liquid by stirring in ingredients such as eggs, flour, butter, or cream until it holds together.

Blackened: Popular Cajun-style cooking method. Seasoned foods are cooked over high heat in a super-heated heavy skillet until charred.

Blanch: To scald, as in vegetables being prepared for freezing; as in almonds so as to remove skins.

Blend: To mix or fold two or more ingredients together to obtain equal distribution throughout the mixture.

Braise: To brown meat in oil or other fat and then cook slowly in liquid. The effect of braising is to tenderize the meat.

Bread: To coat food with crumbs (usually with soft or dry bread crumbs), sometimes seasoned.

Brown: To quickly sauté, broil, or grill either at the beginning or at the end of meal preparation, often to enhance flavor, texture, or eye appeal.

Brush: To use a pastry brush to coat a food such as meat or pastry with melted butter, glaze, or other liquid.

Butterfly: To cut open a food such as pork chops down the center without cutting all the way through, and then spread apart.

Caramelization: Browning sugar over a flame, with or without the addition of some water to aid the process. The temperature range in which sugar caramelizes is approximately 320 to 360 degrees F.

Clarify: To remove impurities from butter or stock by heating the liquid, then straining or skimming it.

Coddle: A cooking method in which foods (such as eggs) are put in separate containers and placed in a pan of simmering water for slow, gentle cooking.

Confit: To slowly cook pieces of meat in their own gently rendered fat.

Core: To remove the inedible center of fruits such as pineapples.

Cream: To beat vegetable shortening, butter, or margarine, with or without sugar, until light and fluffy. This process traps in air bubbles, later used to create height in cookies and cakes.

Crimp: To create a decorative edge on a pie crust. On a double pie crust, this also seals the edges together.

Curd: A custard-like pie or tart filling flavored with juice and zest of citrus fruit, usually lemon, although lime and orange may also be used.

Curdle: To cause semisolid pieces of coagulated protein to develop in food, usually as a result of the addition of an acid substance, or the overheating of milk or egg-based sauces.

Custard: A mixture of beaten egg, milk, and possibly other ingredients such as sweet or savory flavorings, which are cooked with gentle heat, often in a water bath or double boiler. As pie filling, the custard is frequently cooked and chilled before being layered into a baked crust.

Deglaze: To add liquid to a pan in which foods have been fried or roasted, in order to dissolve the caramelized juices stuck to the bottom of the pan.

Dot: To sprinkle food with small bits of an ingredient such as butter to allow for even melting.

Dredge: To sprinkle lightly and evenly with sugar or flour. A dredger has holes pierced on the lid to sprinkle evenly.

Drippings: The liquids left in the bottom of a roasting or frying pan after meat is cooked. Drippings are generally used for gravies and sauces.

Drizzle: To pour a liquid such as a sweet glaze or melted butter in a slow, light trickle over food.

Dust: To sprinkle food lightly with spices, sugar, or flour for a light coating.

Egg Wash: A mixture of beaten eggs (yolks, whites, or whole eggs) with either milk or water. Used to coat cookies and other baked goods to give them a shine when baked.

Emulsion: A mixture of liquids, one being a fat or oil and the other being water based so that tiny globules of one are suspended in the other. This may involve the use of stabilizers, such as egg or custard. Emulsions may be temporary or permanent.

Entrée: A French term that originally referred to the first course of a meal, served after the soup and before the meat courses. In the United States, it refers to the main dish of a meal.

Fillet: To remove the bones from meat or fish for cooking.

Filter: To remove lumps, excess liquid, or impurities by passing through paper or cheesecloth.

Firm-Ball Stage**:** In candy making, the point at which boiling syrup dropped in cold water forms a ball that is compact yet gives slightly to the touch.

Flambé: To ignite a sauce or other liquid so that it flames.

Flan: An open pie filled with sweet or savory ingredients; also, a Spanish dessert of baked custard covered with caramel.

Flute: To create a decorative scalloped or undulating edge on a pie crust or other pastry.

Fricassee: Usually a stew in which the meat is cut up, lightly cooked in butter, and then simmered in liquid until done.

Frizzle: To cook thin slices of meat in hot oil until crisp and slightly curly.

Ganache: A rich chocolate filling or coating made with chocolate, vegetable shortening, and possibly heavy cream. It can coat cakes or cookies, and be used as a filling for truffles.

Glaze: A liquid that gives an item a shiny surface. Examples are fruit jams that have been heated or chocolate thinned with melted vegetable shortening. Also, to cover a food with such a liquid.

Gratin: To bind together or combine food with a liquid such as cream, milk, béchamel sauce, or tomato sauce, in a shallow dish. The mixture is then baked until cooked and set.

Hard-Ball Stage**:** In candy making, the point at which syrup has cooked long enough to form a solid ball in cold water.

Hull (also husk): To remove the leafy parts of soft fruits, such as strawberries or blackberries.

Infusion: To extract flavors by soaking them in liquid heated in a covered pan. The term also refers to the liquid resulting from this process.

Jerk or Jamaican Jerk Seasoning: A dry mixture of various spices such as chilies, thyme, garlic, onions, and cinnamon or cloves used to season meats such as chicken or pork.

Julienne: To cut into long, thin strips.

Jus: The natural juices released by roasting meats.

Larding: To inset strips of fat into pieces of meat, so that the braised meat stays moist and juicy.

Marble: To gently swirl one food into another.

Marinate: To combine food with aromatic ingredients to add flavor.

Meringue: Egg whites beaten until they are stiff, then sweetened. It can be used as the topping for pies or baked as cookies.

Mull: To slowly heat cider with spices and sugar.

Parboil: To partly cook in a boiling liquid.

Peaks: The mounds made in a mixture. For example, egg white that has been whipped to stiffness. Peaks are "stiff" if they stay upright or "soft" if they curl over.

Pesto: A sauce usually made of fresh basil, garlic, olive oil, pine nuts, and cheese. The ingredients are finely chopped and then mixed, uncooked, with pasta. Generally, the term refers to any uncooked sauce made of finely chopped herbs and nuts.

Pipe: To force a semisoft food through a bag (either a pastry bag or a plastic bag with one corner cut off) to decorate food.

Pressure Cooking: To cook using steam trapped under a locked lid to produce high temperatures and achieve fast cooking time.

Purée: To mash or sieve food into a thick liquid.

Ramekin: A small baking dish used for individual servings of sweet and savory dishes.

Reduce: To cook liquids down so that some of the water evaporates.

Refresh: To pour cold water over freshly cooked vegetables to prevent further cooking and to retain color.

Roux: A cooked paste usually made from flour and butter used to thicken sauces.

Sauté: To cook foods quickly in a small amount of oil in a skillet or sauté pan over direct heat.

Scald: To heat a liquid, usually a dairy product, until it almost boils.

Sear: To seal in a meat's juices by cooking it quickly using very high heat.

Seize: To form a thick, lumpy mass when melted (usually applies to chocolate).

Sift: To remove large lumps from a dry ingredient such as flour or confectioners' sugar by passing it through a fine mesh. This process also incorporates air into the ingredients, making them lighter.

Simmer: To cook food in a liquid at a low enough temperature that small bubbles begin to break the surface.

Steam: To cook over boiling water in a covered pan. This method keeps foods' shape, texture, and nutritional value intact better than methods such as boiling.

Steep: To soak dry ingredients (tea leaves, ground coffee, herbs, spices, etc.) in liquid until the flavor is infused into the liquid.

Stewing: To brown small pieces of meat, poultry, or fish, then simmer them with vegetables or other ingredients in enough liquid to cover them, usually in a closed pot on the stove, in the oven, or with a slow cooker.

Thin: To reduce a mixture's thickness with the addition of more liquid.

Truss: To use string, skewers, or pins to hold together a food to maintain its shape while it cooks (usually applied to meat or poultry).

Unleavened: Baked goods that contain no agents to give them volume, such as baking powder, baking soda, or yeast.

Vinaigrette: A general term referring to any sauce made with vinegar, oil, and seasonings.

Zest: The thin, brightly colored outer part of the rind of citrus fruits. It contains volatile oils, used as a flavoring.

Recipe Index of Coconut Delights

Reader Feedback Form

Dear Reader,

We are very interested in what our readers think. Please fill in the form below and return it to:

Whispering Pine Press International, Inc.
c/o Coconut Delights Cookbook
P.O. Box 214, Spokane Valley, WA 99037-0214
Phone: (509) 928-8700 | Fax: (509) 922-9949
Email: sales@whisperingpinepress.com
Publisher Websites: www.WhisperingPinePress.com
www.WhisperingPinePressBookstore.com
Blog: www.WhisperingPinePressBlog.com

Name: ____________________________________

Address: __________________________________

City, St., Zip: ______________________________

Phone/Fax: (____) ______________ / (____) __________

Email: ____________________________________

Comments/Suggestions: ________________________

____

____

____

____

A great deal of care and attention has been exercised in the creation of this book. Designing a great cookbook that is original, fun, and easy to use has been a job that required many hours of diligence, creativity, and research. Although we strive to make this book completely error free, errors and discrepancies may not be completely excluded. If you come across any errors or discrepancies, please make a note of them and send them to our publishing office. We are constantly updating our manuscripts, eliminating errors, and improving quality.

Please contact us at the address above.

About the Cookbook Delights Series

The *Cookbook Delights Series* includes many different topics and themes. If you have a passion for food and wish to know more information about different foods, then this series of cookbooks will be beneficial to you. Each book features a different type of food, such as avocados, strawberries, huckleberries, salmon, vegetarian, lentils, almonds, cherries, coconuts, lemons, and many, many more.

The *Cookbook Delights Series* not only includes cookbooks about individual foods but also includes several holiday-themed cookbooks. Whatever your favorite holiday may be, chances are we have a cookbook with recipes designed with that holiday in mind. Some examples include *Halloween Delights, Thanksgiving Delights, Christmas Delights, Valentine Delights, Mother's Day Delights, St. Patrick's Day Delights,* and *Easter Delights.*

Each cookbook is designed for easy use and is organized into alphabetical sections. Over 250 recipes are included along with other interesting facts, folklore, and history of the featured food or theme. Each book comes with a beautiful full-color cover, ordering information, and a list of other upcoming books in the series.

Note cards, bookmarks, and a daily journal have been printed and are available to go along with each cookbook. You may view the entire line of cookbooks, journals, cards, posters, puzzles, and bookmarks by visiting our websites at www.CookbookDelights.net and www.CoconutDelights.com, or you can email us with your questions and your comments to: sales@whisperingpinepress.com.

Please ask your local bookstore to carry these sets of books.

To order, please contact:

Whispering Pine Press International, Inc.

c/o Coconut Delights Cookbook

P.O. Box 214, Spokane Valley, WA 99037-0214

Phone: (509) 928-8700 | Fax: (509) 922-9949

Email: sales@whisperingpinepress.com

Publisher Websites: www.WhisperingPinePress.com

www.WhisperingPinePressBookstore.com

Blog: www.WhisperingPinePressBlog.com

SAN 253-200X

We Invite You to Join the Whispering Pine Press, Inc., Book Club!

Whispering Pine Press International, Inc.
c/o Coconut Delights Cookbook
P.O. Box 214, Spokane Valley, WA 99037-0214
Phone: (509) 928-8700 | Fax: (509) 922-9949
Email: sales@whisperingpinepress.com
Publisher Websites: www.WhisperingPinePress.com
www.WhisperingPinePressBookstore.com
Blog: www.WhisperingPinePressBlog.com

Buy 11 books and get the next one free, based on the average price of the first eleven purchased.

How the club works:

Simply use the order form below and order books from our catalog. You can buy just one at a time or all eleven at once. After the first eleven books are purchased, the next one is free. Please add shipping and handling as listed on this form. There are no purchase requirements at any time during your membership. Free book credit is based on the average price of the first eleven books purchased.

Join today! Pick your books and mail in the form today!

Yes! I want to join the Whispering Pine Press, Inc., Book Club! Enroll me and send the books indicated below.

Title Price

1. ____________________
2. ____________________
3. ____________________
4. ____________________
5. ____________________
6. ____________________
7. ____________________
8. ____________________
9. ____________________
10. ____________________
11. ____________________

Free Book Title: ____________________

Free Book Price: __________ Avg. Price: __________ Total Price: ________

Credit for the free book is based on the average price of the first 11 books purchased.

(Circle one) Check | Visa | MasterCard | Discover | American Express

Credit Card #: ____________________ Expiration Date: ________

Name: ____________________

Address: ____________________

City: ____________________ State: ________ Country: ________

Zip/Postal: ____________________ Phone: (______) ____________

Email: ____________________

Signature ____________________

Whispering Pine Press, Inc.
Fundraising Opportunities

Fundraising cookbooks are proven moneymakers and great keepsake providers for your group. Whispering Pine Press, Inc., offers a very special personalized cookbook fundraising program that encourages success to organizations all across the USA.

Our prices are competitive and fair. Currently, we offer a special of 100 books with many free features and excellent customer service. Any purchase you make is guaranteed first-rate.

Flexibility is not a problem. If you have special needs, we guarantee our cooperation in meeting each of them. Our goal is to create a cookbook that goes beyond your expectations. We have the confidence and a record that promises continual success.

Another great fundraising program is the *Cookbook Delights Series* Program. With cookbook orders of 50 copies or more, your organization receives a huge discount, making for a prompt and lucrative solution.

We also specialize in assisting group fundraising – Christian, community, nonprofit, and academic among them. If you are struggling for a new idea, something that will enhance your success and broaden your appeal, Whispering Pine Press, Inc., can help.

For more information, write, phone, or fax to:

Whispering Pine Press International, Inc.
P.O. Box 214
Spokane Valley, WA 99037-0214
Phone: (509) 928-8700 | Fax: (509) 922-9949
Email: sales@whisperingpinepress.com
Publisher Websites: www.WhisperingPinePress.com
www.WhisperingPinePressBookstore.com
Blog: www.WhisperingPinePressBlog.com
Book Website: www.CoconutDelights.com
SAN 253-200X

Personalized and/or Translated Order Form for Any Book by Whispering Pine Press International, Inc.

Dear Readers:

If you or your organization wishes to have this book or any other of our books personalized, we will gladly accommodate your needs. For instance, if you would like to change the names of the characters in a book to the names of the children in your family or Sunday school class, we would be happy to work with you on such a project. We can add more information of your choosing and customize this book especially for your family, group, or organization.

We are also offering an option of translating your book into another language. Please fill out the form below telling us exactly how you would like us to personalize your book.

Please send your request to:

Whispering Pine Press International, Inc.
P.O. Box 214, Spokane Valley, WA 99037-0214
Phone: (509) 928-8700 | Fax: (509) 922-9949
Email: sales@whisperingpinepress.com
Publisher Websites: www.WhisperingPinePress.com
www.WhisperingPinePressBookstore.com
Blog: www.WhisperingPinePressBlog.com

Person/Organization placing request: ______________________
Date__________ Phone: (____) ______________________
Address______________________________ Fax: (____) __________
City____________________ State________ Zip: ______________
Language of the book: ______________________
Please explain your request in detail: ______________________
____
____
____
____
____
____

Coconut Delights Cookbook

A Collection of Coconut Recipes

How to Order

Get your additional copies of this book by returning an order form and your check, money order, or credit card information to:

Whispering Pine Press International, Inc.
P.O. Box 214, Spokane Valley, WA 99037-0214
Phone: (509) 928-8700 | Fax: (509) 922-9949
Email: sales@whisperingpinepress.com
Publisher Websites: www.WhisperingPinePress.com
www.WhisperingPinePressBookstore.com
Blog: www.WhisperingPinePressBlog.com

Customer Name: ______________________________

Address: ______________________________

City, St., Zip: ______________________________

Phone/Fax: ______________________________

Email: ______________________________

- -

Please send me ______ copies of ______________________ ____________ at $________ per copy and $4.95 for shipping and handling per book, plus $2.95 each for additional books. Enclosed is my check, money order, or charge my account for $_____________.

☐ Check ☐ Money Order ☐ Credit Card

(*Circle One*) MasterCard | Discover | Visa | American Express

☐☐☐☐ ☐☐☐☐ ☐☐☐☐ ☐☐☐☐

Expiration Date: ____________________

____
Signature

____
Print Name

Whispering Pine Press International, Inc. Order Form

Gift-wrapping, Autographing, and Inscription

We are proud to offer personal autographing by the author. For a limited time this service is absolutely free! Gift-wrapping is also available for $4.95 per item.

1. Sold To

Name: ____________________
Street/Route: ____________________
____
City: ____________________
State: __________ Zip: __________
Country: ____________________
Gift message: ____________________
____
Email address: ____________________
Daytime Phone: (_ _) _ _ _-_ _ _ _
*Necessary for verifying orders
Home Phone: (_ _) _ _ _-_ _ _ _
Fax: (_ _) _ _ _-_ _ _ _

2. Ship To

☐ Is this a new or corrected address?
☐ Alternative Shipping Address
☐ Mailing Address
Name: ____________________
Address: ____________________
____
City: ____________________
State: __________ Zip: __________
Country: ____________________
Email address: ____________________

3. Items Ordered

ISBN # /Item #	Size	Color	Qty.	Title or Description	Price	Total

4. Method Of Payment

International, Inc. (No Cash or COD's)

☐ Visa ☐ MasterCard ☐ Discover ☐ American Express ☐ Check/Money Order

Please make it payable to Whispering Pine Press International, Inc. (No Cash or COD's)

Account Number

Expiration Date
_____/_____
Month Year

☐☐☐☐ ☐☐☐☐ ☐☐☐☐ ☐☐☐☐

Signature____________________
Cardholder's signature

Printed Name____________________
Please print name of cardholder

Address of Cardholder____________________

Subtotal	
Gift wrap $4.95 Each	
For delivery in WA add 8.7% sales tax.	
Shipping See chart at left	
6. Total	

5. Shipping & Handling

Continental US

US Postal Ground: For books please add $4.95 for the first book and $2.95 each for additional books.
All non-book items, add 15% of the Subtotal.
Please allow 1-4 weeks for delivery.
US Postal Air: Please add $15.00 shipping and handling.
Please allow 1-3 days for delivery.
Alaska, Hawaii, and the US Territories By Ship:
Please add 10% shipping and handling (minimum charge $15.00).

Please
By Air: Please add 12% shipping and handling (minimum charge $15.00).
Please allow 2 –6 weeks for delivery.
International By Ship: Please add 10% shipping and handling (minimum charge $15.00).
Please allow 6-12 weeks for delivery.
By Air: Please add 12% shipping and handling (minimum charge $15.00).
Please allow 2-6 weeks for delivery.
FedEx Shipments: Add $5.00 to the above airmail charges for overnight delivery.

Shop Online:
www.whisperingpinepress.com
Fax orders to: (509) 922-9949

Whispering Pine Press International, Inc.
P.O. Box 214
Spokane Valley, WA 99037-0214 USA
Phone: (509) 928-8700 • Fax: (509) 922-9949
Email: sales@whisperingpinepress.com
Website: www.whisperingpinepress.com

Whispering Pine Press International, Inc. Order Form

Gift-wrapping, Autographing, and Inscription

We are proud to offer personal autographing by the author. For a limited time this service is absolutely free!
Gift-wrapping is also available for $4.95 per item.

1. Sold To

Name: ______
Street/Route: ______
______
City: ______
State: ______ Zip: ______
Country: ______
Gift message: ______
______
Email address: ______
Daytime Phone: (_ _ _) _ _ _-_ _ _ _
*Necessary for verifying orders
Home Phone: (_ _ _) _ _ _-_ _ _ _
Fax: (_ _ _) _ _ _-_ _ _ _

2. Ship To

☐ Is this a new or corrected address?
☐ Alternative Shipping Address
☐ Mailing Address
Name: ______
Address: ______
______
City: ______
State: ______ Zip: ______
Country: ______
Email address: ______

3. Items Ordered

ISBN # /Item #	Size	Color	Qty.	Title or Description	Price	Total

4. Method Of Payment

International, Inc. (No Cash or COD's)

☐ Visa ☐ MasterCard ☐ Discover ☐ American Express ☐ Check/Money Order
Please make it payable to Whispering Pine Press International, Inc. (No Cash or COD's)

Account Number
Expiration Date ____/____ Month Year

☐☐☐☐ ☐☐☐☐ ☐☐☐☐ ☐☐☐☐

Signature______
Cardholder's signature

Printed Name______
Please print name of cardholder

Address of Cardholder______

Subtotal	
Gift wrap $4.95 Each	
For delivery in WA add 8.7% sales tax.	
Shipping See chart at left	
6. Total	

5. Shipping & Handling

Continental US
US Postal Ground: For books please add $4.95 for the first book and $2.95 each for additional books.
All non-book items, add 15% of the Subtotal.
Please allow 1-4 weeks for delivery.
US Postal Air: Please add $15.00 shipping and handling.
Please allow 1-3 days for delivery.
Alaska, Hawaii, and the US Territories By Ship:
Please add 10% shipping and handling (minimum charge $15.00).
Please
By Air: Please add 12% shipping and handling (minimum charge $15.00).
Please allow 2 –6 weeks for delivery.
International By Ship: Please add 10% shipping and handling (minimum charge $15.00).
Please allow 6-12 weeks for delivery.
By Air: Please add 12% shipping and handling (minimum charge $15.00).
Please allow 2-6 weeks for delivery.
FedEx Shipments: Add $5.00 to the above airmail charges for overnight delivery.

Shop Online:
www.whisperingpinepress.com
Fax orders to: (509) 922-9949

Whispering Pine Press International, Inc.
P.O. Box 214
Spokane Valley, WA 99037-0214 USA
Phone: (509) 928-8700 • Fax: (509) 922-9949
Email: sales@whisperingpinepress.com
Website: www.whisperingpinepress.com

Current and Future Cookbooks
By Karen Jean Matsko Hood

DELIGHTS SERIES

Almond Delights
Anchovy Delights
Apple Delights
Apricot Delights
Artichoke Delights
Asparagus Delights
Avocado Delights
Banana Delights
Barley Delights
Basil Delights
Bean Delights
Beef Delights
Beer Delights
Beet Delights
Blackberry Delights
Blueberry Delights
Bok Choy Delights
Boysenberry Delights
Brazil Nut Delights
Broccoli Delights
Brussels Sprouts Delights
Buffalo Berry Delights
Butter Delights
Buttermilk Delights
Cabbage Delights
Calamari Delights
Cantaloupe Delights
Caper Delights
Cardamom Delights
Carrot Delights
Cashew Delights
Cauliflower Delights
Celery Delights
Cheese Delights
Cherry Delights
Chestnut Delights
Chicken Delights
Chili Pepper Delights
Chive Delights
Chocolate Delights
Chokecherry Delights
Cilantro Delights
Cinnamon Delights
Clam Delights
Clementine Delights
Coconut Delights
Coffee Delights
Conch Delights
Corn Delights
Cottage Cheese Delights
Crab Delights
Cranberry Delights
Cucumber Delights
Cumin Delights
Curry Delights
Date Delights
Edamame Delights
Egg Delights
Eggplant Delights
Elderberry Delights
Endive Delights
Fennel Delights
Fig Delights
Filbert (Hazelnut) Delights
Fish Delights
Garlic Delights
Ginger Delights
Ginseng Delights
Goji Berry Delights
Grape Delights
Grapefruit Delights
Grapple Delights
Guava Delights
Ham Delights
Hamburger Delights
Herb Delights
Herbal Tea Delights
Honey Delights
Honeyberry Delights
Honeydew Delights
Horseradish Delights
Huckleberry Delights
Jalapeño Delights
Jerusalem Artichoke Delights

Jicama Delights
Kale Delights
Kiwi Delights
Kohlrabi Delights
Lavender Delights
Leek Delights
Lemon Delights
Lentil Delights
Lettuce Delights
Lime Delights
Lingonberry Delights
Lobster Delights
Loganberry Delights
Macadamia Nut Delights
Mango Delights
Marionberry Delights
Milk Delights
Mint Delights
Miso Delights
Mushroom Delights
Mussel Delights
Nectarine Delights
Oatmeal Delights
Olive Delights
Onion Delights
Orange Delights
Oregon Berry Delights
Oyster Delights
Papaya Delights
Parsley Delights
Parsnip Delights
Pea Delights
Peach Delights
Peanut Delights
Pear Delights
Pecan Delights
Pepper Delights
Persimmon Delights
Pine Nut Delights
Pineapple Delights
Pistachio Delights
Plum Delights
Pomegranate Delights
Pomelo Delights
Popcorn Delights
Poppy Seed Delights
Pork Delights
Potato Delights
Prickly Pear Cactus Delights
Prune Delights
Pumpkin Delights
Quince Delights
Quinoa Delights
Radish Delights
Raisin Delights
Raspberry Delights
Rhubarb Delights
Rice Delights
Rose Delights
Rosemary Delights
Rutabaga Delights
Salmon Delights
Salmonberry Delights
Salsify Delights
Savory Delights
Scallop Delights
Seaweed Delights
Serviceberry Delights
Sesame Delights
Shallot Delights
Shrimp Delights
Soybean Delights
Spinach Delights
Squash Delights
Star Fruit Delights
Strawberry Delights
Sunflower Seed Delights
Sweet Potato Delights
Swiss Chard Delights
Tangerine Delights
Tapioca Delights
Tayberry Delights
Tea Delights
Teaberry Delights
Thimbleberry Delights
Tofu Delights
Tomatillo Delights
Tomato Delights
Trout Delights
Truffle Delights
Tuna Delights
Turkey Delights

Turmeric Delights
Turnip Delights
Vanilla Delights
Walnut Delights
Wasabi Delights
Watermelon Delights
Wheat Delights
Wild Rice Delights
Yam Delights
Yogurt Delights
Zucchini Delights

CITY DELIGHTS

Chicago Delights
Coeur d'Alene Delights
Great Falls Delights
Honolulu Delights
Minneapolis Delights
Phoenix Delights
Portland Delights
Sandpoint Delights
Scottsdale Delights
Seattle Delights
Spokane Delights
St. Cloud Delights

FOSTER CARE

Foster Children Cookbook and Activity Book
Foster Children's Favorite Recipes
Holiday Cookbook for Foster Families

GENERAL THEME DELIGHTS

Appetizer Delights
Baby Food Delights
Barbeque Delights
Beer-Making Delights
Beverage Delights
Biscotti Delights
Bisque Delights
Blender Delights
Bread Delights
Bread Maker Delights
Breakfast Delights
Brunch Delights
Cake Delights
Campfire Food Delights
Candy Delights
Canned Food Delights
Cast Iron Delights
Cheesecake Delights
Chili Delights
Chowder Delights
Cocktail Delights
College Cooking Delights
Comfort Food Delights
Cookie Delights
Cooking for One Delights
Cooking for Two Delights
Cracker Delights
Crepe Delights
Crockpot Delights
Dairy Delights
Dehydrated Food Delights
Dessert Delights
Dinner Delights
Dutch Oven Delights
Foil Delights
Fondue Delights
Food Processor Delights
Fried Food Delights
Frozen Food Delights
Fruit Delights
Gelatin Delights
Grilled Delights
Hiking Food Delights
Ice Cream Delights
Juice Delights
Kid's Delights
Kosher Diet Delights
Liqueur-Making Delights
Liqueurs and Spirits Delights
Lunch Delights
Marinade Delights
Microwave Delights
Milk Shake and Malt Delights
Panini Delights
Pasta Delights
Pesto Delights

Phyllo Delights
Pickled Food Delights
Picnic Food Delights
Pizza Delights
Preserved Delights
Pudding and Custard Delights
Quiche Delights
Quick Mix Delights
Rainbow Delights
Salad Delights
Salsa Delights
Sandwich Delights
Sea Vegetable Delights
Seafood Delights
Smoothie Delights
Snack Delights
Soup Delights
Supper Delights
Tart Delights
Torte Delights
Tropical Delights
Vegan Delights
Vegetable Delights
Vegetarian Delights
Vinegar Delights
Wildflower Delights
Wine Delights
Winemaking Delights
Wok Delights

GIFTS-IN-A-JAR SERIES

Beverage Gifts-in-a-Jar
Christmas Gifts-in-a-Jar
Cookie Gifts-in-a-Jar
Gifts-in-a-Jar
Gifts-in-a-Jar Catholic
Gifts-in-a-Jar Christian
Holiday Gifts-in-a-Jar
Soup Gifts-in-a-Jar

HEALTH-RELATED DELIGHTS

Achalasia Diet Delights
Adrenal Health Diet Delights
Anti-Acid Reflux Diet Delights
Anti-Cancer Diet Delights
Anti-Inflammation Diet Delights
Anti-Stress Diet Delights
Arthritis Delights
Bone Health Diet Delights
Diabetic Diet Delights
Diet for Pink Delights
Fibromyalgia Diet Delights
Gluten-Free Diet Delights
Healthy Breath Diet Delights
Healthy Digestion Diet Delights
Healthy Heart Diet Delights
Healthy Skin Diet Delights
Healthy Teeth Diet Delights
High-Fiber Diet Delights
High-Iodine Diet Delights
High-Protein Diet Delights
Immune Health Diet Delights
Kidney Health Diet Delights
Lactose-Free Diet Delights
Liquid Diet Delights
Liver Health Diet Delights
Low-Calorie Diet Delights
Low-Carb Diet Delights
Low-Fat Diet Delights
Low-Sodium Diet Delights
Low-Sugar Diet Delights
Lymphoma Health Support Diet Delights
Multiple Sclerosis Healthy Diet Delights
No Flour No Sugar Diet Delights
Organic Food Delights
pH-Friendly Diet Delights
Pregnancy Diet Delights
Raw Food Diet Delights
Sjögren's Syndrome Diet Delights
Soft Food Diet Delights
Thyroid Health Diet Delights

HOLIDAY DELIGHTS

Christmas Delights
Easter Delights

Father's Day Delights
Fourth of July Delights
Grandparent's Day Delights
Halloween Delights
Hanukkah Delights
Labor Day Weekend Delights
Memorial Day Weekend Delights
Mother's Day Delights
New Year's Delights
St. Patrick's Day Delights
Thanksgiving Delights
Valentine Delights

HOOD AND MATSKO FAMILY FAVORITES

Hood and Matsko Family Appetizers Cookbook
Hood and Matsko Family Beverages Cookbook
Hood and Matsko Family Breads and Rolls Cookbook
Hood and Matsko Family Breakfasts Cookbook
Hood and Matsko Family Cakes Cookbook
Hood and Matsko Family Candies Cookbook
Hood and Matsko Family Casseroles Cookbook
Hood and Matsko Family Cookies Cookbook
Hood and Matsko Family Desserts Cookbook
Hood and Matsko Family Dressings, Sauces, and Condiments Cookbook
Hood and Matsko Family Ethnic Cookbook
Hood and Matsko Family Jams, Jellies, Syrups, Preserves, and Conserves
Hood and Matsko Family Main Dishes Cookbook
Hood and Matsko Family, Pies Cookbook
Hood and Matsko Family Preserving Cookbook
Hood and Matsko Family Salads and Salad Dressings
Hood and Matsko Family Side Dishes Cookbook
Hood and Matsko Family Vegetable Cookbook
Hood and Matsko Family, Aunt Katherine's Recipe Collection, Vol. I-II
Hood and Matsko Family, Grandma Bert's Recipe Collection, Vol. I-IV

HOOD AND MATSKO FAMILY HOLIDAY

Hood and Matsko Family Favorite Birthday Recipes
Hood and Matsko Family Favorite Christmas Recipes
Hood and Matsko Family Favorite Christmas Sweets
Hood and Matsko Family Easter Cookbook
Hood and Matsko Family Favorite Thanksgiving Recipes

INTERNATIONAL DELIGHTS

African Delights
African American Delights
Australian Delights
Austrian Delights
Brazilian Delights
Canadian Delights
Chilean Delights
Chinese Delights
Czechoslovakian Delights
English Delights
Ethiopian Delights
Fijian Delights
French Delights
German Delights
Greek Delights
Hungarian Delights

Icelandic Delights
Indian Delights
Irish Delights
Italian Delights
Korean Delights
Kosovo Delights
Macedonia Republic Delights
Mexican Delights
Montenegro Delights
Native American Delights
Polish Delights
Russian Delights
Scottish Delights
Serbian Delights
Slovakian Delights
Slovenian Delights
Sri Lanka Delights
Swedish Delights
Thai Delights
The Netherlands Delights
Yugoslavian Delights
Zambian Delights

REGIONAL DELIGHTS
Glacier National Park Delights
Northwest Regional Delights
Oregon Coast Delights
Schweitzer Mountain Delights
Southwest Regional Delights
Tropical Delights
Washington Wine Country Delights
Wine Delights of Walla Walla Wineries
Yellowstone National Park Delights

SEASONAL DELIGHTS
Autumn Harvest Delights
Spring Harvest Delights
Summer Harvest Delights
Winter Harvest Delights

SPECIAL EVENTS DELIGHTS
Birthday Delights
Coffee Klatch Delights
Super Bowl Delights
Tea Time Delights

STATE DELIGHTS
Alaska Delights
Arizona Delights
Georgia Delights
Hawaii Delights
Idaho Delights
Illinois Delights
Iowa Delights
Louisiana Delights
Minnesota Delights
Montana Delights
North Dakota Delights
Oregon Delights
South Dakota Delights
Texas Delights
Washington Delights

U.S. TERRITORIES DELIGHTS
Cruzan Delights
U.S. Virgin Island Delights

MISCELLANEOUS COOKBOOKS
Getaway Studio Cookbook
The Soup Doctor's Cookbook

BILINGUAL DELIGHTS SERIES
Apple Delights, English-French Edition
Apple Delights, English-Russian Edition
Apple Delights, English-Spanish Edition
Huckleberry Delights, English-French Edition
Huckleberry Delights, English-Russian Edition
Huckleberry Delights, English-Spanish Edition

CATHOLIC DELIGHTS SERIES

Apple Delights Catholic
Coffee Delights Catholic
Easter Delights Catholic
Huckleberry Delights Catholic
Tea Delights Catholic

CATHOLIC BILINGUAL DELIGHTS SERIES

Apple Delights Catholic, English-French Edition
Apple Delights Catholic, English-Russian Edition
Apple Delights Catholic, English-Spanish Edition
Huckleberry Delights Catholic, English-Spanish Edition

CHRISTIAN DELIGHTS SERIES

Apple Delights Christian
Coffee Delights Christian
Easter Delights Christian
Huckleberry Delights Christian
Tea Delights Christian

CHRISTIAN BILINGUAL DELIGHTS SERIES

Apple Delights Christian, English-French Edition
Apple Delights Christian, English-Russian Edition
Apple Delights Christian, English-Spanish Edition
Huckleberry Delights Christian, English-Spanish Edition

FUNDRAISING COOKBOOKS

Ask about our fundraising cookbooks to help raise funds for your organization.

Many of the above books are also available in bilingual versions. Please contact Whispering Pine Press International, Inc., for details.

The above list of books is not all-inclusive. For a complete list please visit our website or contact us at:

Whispering Pine Press International, Inc.
Your Northwest Book Publishing Company
P.O. Box 214
Spokane Valley, WA 99037-0214 USA
Phone: (509) 928-8700 | Fax: (509) 922-9949
Email: sales@whisperingpinepress.com
Publisher Websites: www.WhisperingPinePress.com
www.WhisperingPinePressBookstore.com
Blog: www.WhisperingPinePressBlog.com

About the Author and Cook

Karen Jean Matsko Hood has always enjoyed cooking, baking, and experimenting with recipes. At this time Hood is working to complete a series of cookbooks that blends her skills and experience in cooking and entertaining. Hood entertains large groups of people and especially enjoys designing creative menus with holiday, international, ethnic, and regional themes.

Hood is publishing a cookbook series entitled the *Cookbook Delights Series,* in which each cookbook emphasizes a different food ingredient or theme. The first cookbook in the series is *Apple Delights Cookbook.* Hood is working to complete another series of cookbooks titled *Hood and Matsko Family Cookbooks,* which includes many recipes handed down from her family heritage and others that have emerged from more current family traditions. She has been invited to speak on talk radio shows on various topics, and favorite recipes from her cookbooks have been prepared on local television programs.

Hood was born and raised in Great Falls, Montana. As an undergraduate, she attended the College of St. Benedict in St. Joseph, Minnesota, and St. John's University in Collegeville, Minnesota. She attended the University of Great Falls in Great Falls, Montana. Hood received a B.S. Degree in Natural Science from the College of St. Benedict and minored in both Psychology and Secondary Education. Upon her graduation, Hood and her husband taught science and math on the island of St. Croix in the U.S. Virgin Islands. Hood has completed postgraduate classes at the University of Iowa in Iowa City, Iowa. In May 2001, she completed her Master's Degree in Pastoral Ministry at Gonzaga University in Spokane, Washington. She has taken postgraduate classes at Lewis and Clark College on the North Idaho college campus in Coeur d'Alene, Idaho, Taylor University in Fort Wayne, Indiana, Spokane Falls Community College, Spokane Community College, Washington State University, University of Washington, and Eastern Washington University. Hood is working on research projects to complete her Ph.D. in Leadership Studies at Gonzaga University in Spokane, Washington.

Hood resides in Greenacres, Washington, along with her husband, many of her sixteen children, and foster children. Her interests include writing, research, and teaching. She previously has volunteered as a court advocate in the Spokane juvenile court system for abused and neglected children. Hood is a literary advocate for youth and adults. Her hobbies include cooking, baking, collecting, photography, indoor and outdoor gardening, farming, and the cultivation of unusual flowering plants and orchids. She enjoys raising several specialty breeds of animals including Babydoll Southdown, Friesen, and Icelandic sheep, Icelandic horses, bichons frisés, cockapoos, Icelandic sheepdogs, a Newfoundland, a Rottweiler, a variety of Nubian and fainting goats, and a few rescue cats. Hood also enjoys bird-watching and finds all aspects of nature precious.

She demonstrates a passionate appreciation of the environment and a respect for all life. She also invites you to visit her websites:

www.KarenJeanMatskoHood.com
www.KarenJeanMatskoHoodBookstore.com
www.KarenJeanMatskoHoodBlog.com
www.KarensKidsBooks.com
www.KarensTeenBooks.com

www.HoodFamilyBlog.com
www.HoodFamily.com

Author's Social Media

Please Follow the Author on **Twitter:** @KarenJeanHood
Friend her on **Facebook:** Karen Jean Matsko Hood Author Fan Page
Google Plus Profile: Karen Jean Matsko Hood
Pinterest.com/KarenJMHood

www.ingramcontent.com/pod-product-compliance
Lightning Source LLC
LaVergne TN
LVHW091642100826
845152LV00006B/129/J

* 9 7 8 1 5 9 4 3 4 2 9 3 6 *